CHEMISTRY RESEARCH AND APPLICATIONS

FORMALDEHYDE

SYNTHESIS, APPLICATIONS AND POTENTIAL HEALTH EFFECTS

CHEMISTRY RESEARCH AND APPLICATIONS

Additional books in this series can be found on Nova's website under the Series tab.

Additional e-books in this series can be found on Nova's website under the e-book tab.

CHEMISTRY RESEARCH AND APPLICATIONS

FORMALDEHYDE

SYNTHESIS, APPLICATIONS AND POTENTIAL HEALTH EFFECTS

AMY PATTON
EDITOR

nova publishers
New York

Library of Congress Cataloging-in-Publication Data

Formaldehyde : synthesis, applications, and potential health effects / Amy Patton, editor.
pages cm. -- (Chemistry research and applications)
Includes index.
ISBN 978-1-63482-412-5 (hardcover)
1. Formaldehyde. 2. Formaldehyde--Environmental aspects. 3. Formaldehyde--Toxicology. I. Patton, Amy, 1972- editor.
TP248.F6F68 2015
615.9'5136--dc23
2015007251

Published by Nova Science Publishers, Inc. † New York

CONTENTS

PREFACE

Formaldehyde (FA) is a highly reactive compound with a toxic effect on all organisms due to non-specific interactions with proteins and nucleic acids. This substance is a hazardous air pollutant, and prolonged exposure to FA may cause serious health effects. This book discusses the synthesis of formaldehyde, its applications and potential health effects.

Chapter 1 – Formaldehyde (FA) is a high-volume production chemical with an annual production and consumption of approximately 29 million tonnes worldwide. It is a building-block of numerous compounds that are used in a wide array of materials and products decisive in everyday life. Given its economic importance and widespread use, many people are exposed to FA both environmentally and/or occupationally. For the general population the major sources of FA exposure include combustion sources, tobacco smoke, off-gassing from numerous construction and home-furnishing products, and off-gassing from consumer goods. The most significant source of indoor contamination is the release from newly home furnishings, primarily from wood pressed products made with urea-formaldehyde resins. These emissions are able to cause FA-exposure related symptoms in residents or occupants especially irritation of the eyes and upper airways. The most known example has occurred in USA, involving hurricane victims and government-provided FEMA (Federal Emergency Management Agency) trailers. However, exposure levels are much higher in the workplace. Listed, since 2004, by IARC (International Agency for Research on Cancer) as *human carcinogen*, FA status was recently revised by USA government who reclassified this compound *as known to be a human carcinogen*. Both reclassifications are based on sufficient evidence of carcinogenicity from studies in humans, supporting studies on mechanisms of carcinogenesis and experimental evidence in animals. Epidemiological studies demonstrated a causal relationship between exposure to FA and cancer. Consisting findings of increased risks of certain types of rare cancers were found among workers with higher measures of exposure to FA (exposure level or duration). In the last decade a large number of toxicological studies were published about FA. FA's genotoxicity is confirmed in a variety of experimental systems ranging from bacteria to rodents. Although these positive findings may provide a basis for extrapolation to humans, the cytogenetic assays in humans have been conflicting with both positive and negative outcomes. The aim of this review is to give a comprehensive knowledge of the principal sources of human exposure to FA. Additionally, an integrated perspective of the principal findings regarding the effects of FA in human health will also be given with an overview of the major findings from epidemiological and toxicological studies.

Chapter 2 – Carbon gels are porous materials obtained from the polymerization of resorcinol with formaldehyde in the presence of a solvent. In this reaction, formaldehyde is the reagent responsible for the crosslinking that takes place during the polymerization and, hence, it plays an important role in determining the final porous structure of these materials. This chapter goes deeply into the way in which formaldehyde modifies the final porous properties of carbon gels and how its influence is affected by other variables involved in the synthesis process. In general terms, an increase in the concentration of formaldehyde shifts the chemical equilibrium towards the products, thereby increasing the reaction rate. This implies that clusters are formed and grow faster. On the other hand, an excess of formaldehyde also leads to more interconnections between clusters, resulting in a materials with highly branched structures. However, the effect of formaldehyde may be modified by the effect of other reagents. For example, an increase in the amount of solvent hinders the crosslinkages produced during the reaction. Similarly, the initial pH of the precursor solution also influences the effect of formaldehyde, since acidic conditions favour the addition of formaldehyde among the resorcinol anions. The use of surfactants also affects the normal course of the sol-gel reaction. Commonly, these additives are either ionic or non-ionic and, depending on their chemical load, their effect on the formation of the porous structure will be different. Generally, anionic surfactants create repulsive forces between resorcinol anions while cationic surfactants generate attractive forces. Depending on the method of preparation of the precursor solution, non-ionic surfactants are either incorporated into the structure of the gel or remain as micelles, leaving voids when the surfactant is removed. Generally, the formaldehyde used in the synthesis of carbon gels (37 wt. % in water) is stabilized by methanol. The role of the methanol is to prevent the polymerization of formaldehyde by forming hemiacetal groups. The concentration of methanol in the solution depends largely on the manufacturer and is usually set between 0.6 and 15 wt.%. This variation in the concentration of methanol also affects the influence of formaldehyde on the polymerization reaction of carbon gels since the formaldehyde needs to be in its free form in order to react with resorcinol. Solutions that contain low concentrations of methanol contain larger amounts of free formaldehyde molecules and therefore, materials with greater porosity are obtained.

Chapter 3 – Worldwide formaldehyde is manipulated with diverse usage properties, since industrial purposes to health laboratory objectives, representing the economic importance of this chemical agent. Therefore, many people are exposed to formaldehyde environmentally and/or occupationally. Considering the latter, there was recommended occupational exposure limits based on threshold mechanisms, limit values and indoor guidelines. Formaldehyde is classified by the International Agency for Cancer Research (IARC) as carcinogenic to humans (group 1), since a wide range of epidemiological studies in occupational exposure settings have suggested possible links between the concentration and duration of exposure and elevated risks of nasopharyngeal cancer, and others cancers, and more recently, with leukemia. Although there are different classifications, such as U.S. EPA that classified formaldehyde as a B1 compound, probable human carcinogen under the conditions of unusually high or prolonged exposure, on basis of limited evidence in humans but with sufficient evidence in animals. Formaldehyde genotoxicity is well-known, being a direct-acting genotoxic compound positively associated for almost all genetic endpoints evaluated in bacteria, yeast, fungi, plants, insects, nematodes, and cultured mammalian cells. There are many human biomonitoring studies that associate formaldehyde occupational exposure to genomic instability, and consequently possible health effects. Besides the link with cancer, also other pathologies and symptoms are associated with formaldehyde exposure, namely respiratory disorders such as asthma, and allergic contact dermatitis. Nowadays, there are efforts to reduce formaldehyde exposure, namely indoor. Europe and United States developed more strict regulation regarding formaldehyde emissions from materials containing this agent. Despite the regulations and restrictions, formaldehyde still continues to be difficult to eliminate or substitute, being biomonitoring an important tool to control possible future health effects.

Chapter 4 – Formaldehyde has recently been placed at the forefront of Alzheimer's disease research. The reasons for this impetus is multi-fold and because the properties of formaldehyde, a planar polar single carbon aldehyde, have strong relevance to the disease. Alzheimer's disease is an age-related neurodegenerative dementia that presents with cognitive decline and memory impairments and is diagnosed with neuritic plaques, neurofibrillary tangles and neuronal atrophy. First, formaldehyde has been found to be an age-related factor whose concentrations in blood and cerebral spinal fluid increase with age and has been found to be higher in Alzheimer's patients post-mortem. Formaldehyde and methanol, a one-carbon precursor of formaldehyde, have been found to cause memory impairments in laboratory animals. Moreover, formaldehyde is a known protein cross-linker that has been shown to cause the aggregation of both amyloid peptides and tau protein as well as tau protein phosphorylation *in vitro*. These properties are central to the two main pathological hallmarks of Alzheimer's disease: the amyloid plaque and neurofibrillary tangle. As such, chronic methanol exposure has been found to induce all of these hallmarks, including memory impairments, in rhesus monkeys. While these findings have implications related to exogenous exposure to both methanol and formaldehyde, there are a number of indications that endogenous metabolic processes may also contribute formaldehyde to the pathological state. Most notably is the condition of hyperhomocysteremia in Alzheimer's disease. Hyperhomocysteremia represents an imbalance of the one-carbon metabolic cycle that favors the metabolite homocysteine, while potentially contributing free one-carbon moieties like formaldehyde to the blood or cerebral spinal fluid. However, no correlation studies between these two related metabolites and Alzheimer's disease have been investigated to date. Nonetheless, formaldehyde offers a unique chemical toxicity that has the potential to explain a large gambit of the pathological progression of the disease; from age-related progression, memory impairments and genetic risk factors. Although there is considerable implications of formaldehyde, such as in the development of novel therapies for the treatment of Alzheimer's disease, very little research and focus have been given to one-carbon toxicity in the past. This brief review serves to focus related research and raise awareness of formaldehyde in Alzheimer's disease pathology so that it may receive better consideration in the future.

Chapter 5 – VOCs exist widely in both indoor & outdoor environment, and have been known as the most hazardous materials which affect public health and welfare owing to their toxicity potential, carcinogenicity and stability. VOCs are recognized as causative agents of the sick-building syndrome, in which formaldehyde (HCHO) is one of them. HCHO has been classified as a suspected carcinogen. HCHO is known to cause nasal tumors, irritation of the mucous membranes of the eyes and respiratory tract, and skin irritation. Because of the increasing concern on HCHO in the indoor environment, the abatement of HCHO is of significant practical interest at low temperature, especially at room temperature. The conventional technique to remove VOCs is activated-carbon adsorption method, which needs to replace and dispose the adsorbent. The disposal of activated-carbon will cause the next environmental problem. The alternative methods are ozone oxidation method, catalytic combustion method, non-thermal plasma method, etc.. The ozone oxidation and the non-thermal plasma methods involve NOx emission and high initial cost. The catalytic combustion method requires a heat source, hence it is a highly energy-wasting process when VOC concentrations are low. In this chapter, two kinds of HCHO decomposition technology were investigated, that is, photo-catalytic degradation and non-thermal plasma technology. Photo-catalytic degradation of VOCs on UV-illuminated titanium dioxide (TiO_2) is proposed as an alternative advanced oxidation process for the purification of water and air. Non-thermal plasma (NTP) has great industrial potential and has been applied to the decomposition of VOCs. At the same time, NTP technology has disadvantages such as low energy efficiency and undesirable byproducts such as NOx. So the combination of NTP with photocatalyst technology was used in the authors' investigation. Nano-structured TiO_2, Ag/TiO_2 and Ce/TiO_2 thin films were prepared by sol-gel method at room temperature and used in the photo-catalyst reactor for HCHO decomposition. The effects of doped $Ag/Ce/TiO_2$, relative humidity, oxygen concentration, initial HCHO concentration, UV light wave length, TiO_2 amount on HCHO decomposition were investigated. At the same time, destruction of formaldehyde by means of $NaNO_2$ ferro-electric packed bed dielectric barrier discharge plasma in a coaxial cylindrical reactor was carried out at atmospheric pressure and room temperature. The difference among four kinds of $NaNO_2$ ferro-electric reactors was compared in terms of specific energy density (SED), energy yield (EY), and HCHO decomposition. Three kinds of catalysts, that is, TiO_2, $BaTiO_3$, or TiO_2 & $BaTiO_3$ were tested for HCHO removal. Effect of catalyst carrier, catalysts, and catalyst amount on specific energy density (SED), energy yield (EY) and HCHO removal efficiency were investigated. In addition, the by-products were detected and the mechanism of HCHO degradation was performed.

Chapter 6 – Formaldehyde (FA) is a highly reactive compound with a toxic effect on all organisms due to non-specific interactions with proteins and nucleic acids. This substance is a hazardous air pollutant, and prolonged exposure to FA may cause serious health effects. This chapter describes approaches for detection and eliminating FA in several real and model sources. The cells of the thermotolerant methylotrophic yeast *Hansenula (Ogatae) polymorpha* as well as FA-oxidizing enzymes isolated from these cells were employed as catalytic elements in the development of enzymatic methods for FA detection by analytical kits and biosensors as well as for degradation in bioreactors. Recombinant yeast strains, originating from NCYC 495 and CBS 4732 *H. polymorpha* strains resistent to elevated concentrations of FA in the medium and overproducing a homologous NAD- and glutathione-dependent formaldehyde dehydrogenase (FdDH), were constructed. Optimal cultivation conditions for the highest yield of FdDH as well as formaldehyde reductase (FR) were found. Alcohol oxidase (AOX) was isolated from the mutant strain *H. polymorpha* C-105 (*gcr1 catX*) which is defective in glucose repression of AOX synthesis and lacks catalase. Simple schemes for FdDH, FR and AOX isolation and chromatographic purification from the recombinant overproducers were proposed and the physico-chemical and catalytic properties of the purified enzymes were studied. Enzymatic methods and analytical kits for FA assay based on glutathione-dependent FdDH and AOX were developed. The purified FdDH and AOX preparations, as well as yeast cells overproducing these enzymes, were used for the construction of FA-selective enzyme-based and microbial electrochemical biosensors. A laboratory prototype of a FA analyzer with a catalytic chemosensing element and a simple low-cost FA-sensitive biosensor for semi-quantitative FA detection were developed. The latter sensor is based on a change in the color of a solution that contains a mixture of AOX, horseradish peroxidase and a chromogen, after exposure to airborne FA. The reliability of the proposed analytical approaches was tested on model solutions and on real samples of FA-containing industrial products, vaccines and wastes. A strong cross-correlation was found between FA content values obtained by the developed methods and routinely used chemical ones. The enzyme preparations and the yeast cells overproducing the corresponding enzymes were used for construction of FA-selective electrochemical biosensors. The purified enzymes and the recombinant and mutant cells served as the catalytic units in cartridges for FA removal from indoor air and from water solutions. Experimental data confirm the possibility of exploiting the developed bioreactors based on crude preparations of AOX or methylotrophic yeast cells, for effective FA oxidation (up to 99.9%) controlled by a coupled FdDH-based biosensor.

Chapter 7 – A mechanism of the initiated nonbranched-chain process of forming 1,2-alkanediols and carbonyl compounds in alcohol–formaldehyde systems is suggested. The quasi-steady-state treatment is used to obtain kinetic equations that can describe the nonmonotonic (with a maximum) dependences of the formation rates of the products on the concentration of free unsolvated formaldehyde. An empirical equation for calculating the concentration of the free unsolvated form of formaldehyde in alcohol–formaldehyde (including water/ethanediol–formaldehyde) systems at various temperatures and total formaldehyde concentrations and an equation for evaluating solvent concentrations in these systems were derived.

Chapter 8 – In this work the authors established influence of mechanical activation on structure and properties oxides of iron and molybdenum. The process of mechanochemical synthesis of molybdates of iron and calcium have been studied, and also their catalytic properties in reaction of partial oxidation of methanol in formaldehyde. It is established that 2% addition in structure of the catalyst of calcium molybdate increases durability of contact weight without essential change of catalytic activity.

In: Formaldehyde
Editor: Amy Patton
ISBN: 978-1-63482-412-5

Chapter 1

FORMALDEHYDE: HUMAN EXPOSURE, METABOLISM AND POTENTIAL HEALTH EFFECTS

Solange Costa* and J. P. Teixeira
National Institute of Health, Portugal

ABSTRACT

Formaldehyde (FA) is a high-volume production chemical with an annual production and consumption of approximately 29 million tonnes worldwide. It is a building-block of numerous compounds that are used in a wide array of materials and products decisive in everyday life. Given its economic importance and widespread use, many people are exposed to FA both environmentally and/or occupationally. For the general population the major sources of FA exposure include combustion sources, tobacco smoke, off-gassing from numerous construction and home-furnishing products, and off-gassing from consumer goods. The most significant source of indoor contamination is the release from newly home furnishings, primarily from wood pressed products made with urea-formaldehyde resins. These emissions are able to cause FA-exposure related symptoms in residents or occupants especially irritation of the eyes and upper airways. The most known example has occurred in USA, involving hurricane victims and government-provided FEMA (Federal Emergency Management Agency) trailers. However, exposure levels are much higher in the workplace. Listed, since 2004, by IARC (International Agency for Research on Cancer) as *human carcinogen*, FA status was recently revised by USA government who reclassified this compound *as known to be a human carcinogen*. Both reclassifications are based on sufficient evidence of carcinogenicity from studies in humans, supporting studies on mechanisms of carcinogenesis and experimental evidence in animals. Epidemiological studies demonstrated a causal relationship between exposure to FA and cancer. Consisting findings of increased risks of certain types of rare cancers were found among workers with higher measures of exposure to FA (exposure level or duration). In the last decade a large number of toxicological studies were published about FA. FA's genotoxicity is confirmed in a variety of experimental systems ranging from bacteria to rodents. Although these positive findings may provide a basis for extrapolation to humans, the cytogenetic assays in humans have been conflicting with both positive and negative outcomes. The aim of this review is to give a comprehensive

* Corresponding author: solange.costa2@gmail.com.

knowledge of the principal sources of human exposure to FA. Additionally, an integrated perspective of the principal findings regarding the effects of FA in human health will also be given with an overview of the major findings from epidemiological and toxicological studies.

1. INTRODUCTION

Formaldehyde (FA) is the most simple of all aliphatic aldehydes. It is a low-molecular weight organic compound, highly reactive, that readily combines with many chemical substances. At room temperature, FA is a flammable and colourless gas with a strong and distinctive odour.

FA is a building-block of numerous compounds that are used in a wide array of materials and products. Due to its versatile properties and inexpensive production, FA is an economically important chemical produced worldwide, with a global annual production and consumption of approximately 29 million tonnes (IHS, 2012). Generally FA is produced near the point of consumption because while it is easy to make, is costly to transport and can develop stability-associated problems during transportation and storage. As a result, international trade is minimal (NTP, 2011) and FA demand is provided by domestic industry.

FA was first synthesised by Aleksandr Butlerov in 1859, but was only in 1868 that it was conclusively identified by August von Hofmann, as the product obtained from passing a mixture of air and methanol over a heated platinum spiral. In fact, this production method laid the foundation for the modern manufacturing process of FA, via methanol oxidation using a metal catalyst (copper, molybdenum alloy, platinum or silver). At the end of the 19th century, FA was already used as a regular medical preservative and embalming agent. By the early 20th century driven by science and technology development, Leo Baekeland used a phenolformaldehyde resin to invent the first synthetic polymeric material, a hard mouldable plastic, precursor of modern synthetic plastics. In 1940s, the first commercial particleboard was produced and a revolution was launched in the furniture and construction industries. Commercial applications continue to grow and today FA has an important role in both medical laboratory procedures and in the production of domestic and industrial goods that have become decisive in everyday life.

FA is usually marketed as a 37 % (by weight) aqueous solution named formalin. Formalin also contains methanol (up to 15%) and other substances (several hundred milligrams per litre) that act as stabilisers. FA is also sold, to a lesser extent, in its solid form, either as a polymer, paraformaldehyde, or as a cyclic trimer, 1,3,5-trioxane.

In many instances, because FA is a long-established material with versatile properties, few compounds can replace it without compromising quality, performance and cost, which hinders its substitution to both producers and users.

2. HUMAN EXPOSURE

Given its economic importance and widespread use, many people are exposed to FA both environmentally and occupationally. The major route of exposure to FA is inhalation (IARC, 2006). Although environmental exposure typically occurs at much lower levels than occupational exposure, people are exposed to FA at some concentration in their daily lives. FA is also a naturally occurring biological compound present in all cells, tissues and body fluids (IARC, 2006).

According to WHO (2000) the estimates for the daily intake of FA in ambient air can vary between 0.002 and 0.04 mg/day mean values. On the other hand, for indoor air the average intake levels can differ between 0.3 and 0.6 mg/day. However, indoor air concentrations can be substantially increased with tobacco smoking, rising up to 0.5 to 3.5 mg/day. WHO also predicts that the mean daily intake of FA from occupational exposure is up to 8 mg/day, being variable according to the work in question.

2.1. Environmental

FA is ubiquitous in the environment and has been detected in indoor and oudoor air, soil, food, surface water, groundwater, treated and bottled water (NTP, 2011).

FA is formed primarily by the combustion of organic materials and by a variety of natural and anthropogenic activities.

Combustion sources include automobiles and other internal combustion engines, power plants, waste incinerators, refineries, wood stoves, forest fires and tobacco smoking. FA is also formed in the early stages of residual plant decomposition in the soil.

Secondary formation of FA occurs in the atmosphere through the photochemical oxidation of natural and anthropogenic hydrocarbon pollutants present in the air. In some instances, given the diversity and abundance of FA precursors in urban air, secondary production may exceed direct air emissions (IPCS, 2002).

For the general population the major sources of FA exposure include combustion sources, tobacco smoke, off-gassing from numerous construction and home-furnishing products, and off-gassing from consumer goods.

FA levels in indoor air are often higher by one order of magnitude or more than those outdoors (IARC, 2006). The most significant source of indoor contamination is the release from newly home furnishings, primarily from wood pressed products made with urea-formaldehyde resins. These emissions are able to cause FA-exposure related symptoms in residents or occupants especially irritation of the eyes and upper airways (Salthammer et al., 2010). The most known example has occurred in the US, involving hurricane victims and government-provided FEMA (Federal Emergency Management Agency) trailers. Numerous victims of Hurricanes Katrina and Rita suffered adverse health effects (respiratory and other symptoms) as a result of being housed in FEMA trailers containing levels of FA 2-fold higher than the reference level. The FA was emitted from the materials used to build the mobile homes (CDC, 2008).

The general population can also be exposed to FA from consumer goods containing FA and from ingestion of food and water (IARC, 2006). Some of the products that often contain FA-releasers are cosmetics, household cleaning agents, soaps, shampoos, paints and lacquers (IPCS, 2002).

Food and water contain measureable concentrations of FA, but the significance of ingestion as a source of FA exposure is not considered relevant (IPCS, 2002; NTP, 2011). In food FA can be present either naturally (e.g., fruit, vegetables, shellfish) or as result of inadvertent contamination. FA-hydrated form, methylene glycol, can be found in surface water or groundwater but usually it is biodegraded within 1 to 15 days (IPCS, 2002). In treated and drinking water FA is formed primarily through oxidation of organic matter during ozonation or chlorination treatment (NTP, 2011), yet its concentration is usually less than 100 μg/L (IPCS, 2002).

2.2. Occupational

The major source of exposure to airborne FA occurs in the work environment. Occupational exposure to FA is transversal to a wide variety of professions and industries.

In the early 1990s it was estimated that approximately 1 million workers in the EU and 2 million in the US were exposed to FA (FIOH, 1999; NTP, 2011). Although no current data is available in the literature, given that FA-world demand has grown over the last decades, it is safe to speculate that the number of individuals occupationally exposed to FA is now higher.

In occupational settings, FA occurs mainly as a gas, hence the major route of exposure is inhalation. Nevertheless, workers may also be exposed through dermal contact (formalin solutions) or by inhalation of FA-particulates such as powdered resins or wood dust (NTP, 2011).

Given its economic importance and wide spread use, occupational exposure to FA involves not only individuals employed in the direct manufacture of FA and products containing it, but also those using the products. Overall, the exposure to FA occurs in three main work-related scenarios:

- production and use of formalin in chemical industry (e.g., synthesis of resins) and in medical laboratories as a preservative and disinfectant agent.
- release from FA-based resins in which it is present as a residue and/or through their hydrolysis and decomposition by heat, during the manufacture of wood products, textiles and plastics.
- the pyrolysis or combustion of organic matter, e.g., in engine exhaust gases or during firefighting.

Another workplace circumstance that in the last few years has raised some concern is the FA emission from hair straightening products during hair treatment in hairdresser salons; although as shown by Pierce et al. (2011) this appeared to be limited to some specific products.

The occupational groups considered to be more at risk, either by duration or level of exposure, are industrial workers from chemical industries and plywood factories, embalmers and pathology and anatomy laboratory workers (IARC, 2006).

According to IARC (2006) the highest exposures to this aldehyde were measured in the past in industrial settings (e.g., varnishing of furniture and wooden floors, garment industry). However, recent legislation (e.g., resins that release less FA) and improved ventilation have resulted in decreased levels of exposure in many industrial workplaces.

Exposures to levels of up to 3 parts per million (ppm) have been reported for embalmers and for hospital laboratory professionals, such as anatomy and pathology workers (IARC, 2006). A recent study assessing the airborne exposure to FA in an industrial unit and anatomy pathology laboratory reveal higher levels of FA in the lab compared to the industrial setting (Viegas et al., 2010).

Furthermore, numerous epidemiological studies indicate a significant association between cancer risk and occupational exposure to FA (Hauptmann et al., 2004; Zhang et al., 2009). In fact, a recent study reported that the estimated cancer risk of laboratory technicians exposed to FA was 20% higher than the general population (Pilidis et al., 2009).

2.2.1. Regulation and Guidelines

The basis for the prevention and control of health risks due to occupational exposure to chemicals are guidelines and standards. Their primary aim is to protect human health from adverse effects of indoor exposure to air pollution and to eliminate or minimise exposure to agents that are or are likely to be hazardous. The guidelines are based on a comprehensive review and evaluation of the accumulated scientific knowledge available at the time of their development and often include expert judgment. They have the character of recommendations and may serve as scientific basis to assist policy-makers and regulatory authorities in setting legally binding limits, the standards. Other factors beyond the toxicological and epidemiological data are taken into account in setting regulatory standards. These factors include environmental, social, cultural and economic conditions, as well as technical feasibility. Nevertheless, governments and regulatory authorities may adopt the recommended limits as legally enforceable standards.

Whether it is guideline or standard, they represent the highest concentration in the air of a chemical under which it is believed that nearly all workers may be repeatedly exposed, day after day, over a working lifetime, without adverse health effects. The *nearly all* refers to individual susceptibility (genetic predisposition), social-demographic factors (age, gender, ethnicity, lifestyle) or others (previous exposures) that may induce an adverse biological response to a particular chemical at a concentration equal or lower than the limit level (ACGIH, 2013).

Three categories of occupational exposure limits (OELs) are usually used to classify the exposure: i) time-weighted average (TWA), concentration that should never be exceeded during workday (8h) or workweek (40h); ii) short-term exposure limit (STEL), concentration that should never be exceeded in 15 min exposure at any time during a workday; iii) ceiling (C), concentration that should never be exceeded during any part of working exposure (direct value). For some substances, a TWA alone or with a STEL is relevant, for others only a C-level is applicable, it depends on the toxicological properties of the chemical and nature of the work.

OELs are express in parts per million (ppm) or mg/m^3 for compounds that normally exist as a gas or vapour at room temperature and pressure.

A "healthy" workplace environment also includes providing instructions, procedures, training and supervision to encourage people to work safely and responsibly. When engineering controls or safe systems are not feasible or effective to reduce the risk to acceptable levels the use of personal protective equipment (PPE) as a control measure may aid to prevent the exposure and thus minimise the risk. Nevertheless, PPE does not eliminate/control the hazard at source and therefore if the equipment fails or if it is not suitable for the chemical in question the workers will not be protected.

Occupational exposure standards and guidelines for airborne-FA set by countries or major international organisations are listed in Table I. The exposure limits recommended by the European Scientific Committee were also included (SCOEL, 2008).

FA-air concentration in workplace is legally regulated all over the world with defined TWA, STEL and C-levels of exposure ranging all three from 0.3 to 2 ppm (IARC, 2006).

Over the time the expansion of scientific knowledge, production of new studies and the updating of evaluation criteria have led to the global review of the FA's occupational exposure limits (OELS) resulting in a reduction of the limit levels (Binetti et al., 2006). Nevertheless, although the number of scientific data on FA toxicity and potential cancer risk has increased in the last decade, the OELs in most countries remain the same (in some the values are not revised since the early 1990s, e.g., US, France), which impairs the establishment of more rigorous protective measures of the worker´s health. Thus, the review by policy-makers of the exposure limit values in light of recent implication of FA carcinogenicity is essential to minimise the risk.

Table 1. Occupational exposure limits for airborne formaldehyde (standards and guidelines

Country/Organisations	OELs (category)	*References*
China	0.4 ppm (C)	*Tang et al. (2009)*
European Commission*		
SCOEL	0.2 ppm (TWA) 0.4 ppm (STEL)	*SCOEL (2008)*
France	0.5 ppm (TWA) 1 ppm (STEL)	*AFSSET (2008)*
Germany	0.3 ppm (TWA) 0.6 ppm (STEL) 1 ppm (C)	*IARC (2006)*
Norway	0.5 ppm (TWA)	*IARC (2006)*
Portugal	0.3 ppm (C)	*IPQ (2007)*
Spain	0.3 ppm (C)	*IARC (2006)*
United Kingdom	2 ppm (TWA) 2 ppm (STEL)	*HSE (2011)*
United States of America		
ACGIH	0.3 (C)	*ACGIH (2013)*
OSHA	0.75 ppm (TWA) 2 ppm (STEL)	*OSHA (2011)*

* Values recommended by the Scientific Committee on Occupational Exposure Limit Values (SCOEL). C, ceiling; TWA, time-weighted average; STEL, short-term exposure limit.

3. TOXICOKINETIC AND METABOLISM

The toxicokinetic behavior of FA has been the subject of several studies in a number of mammal species, and although there may be quantitative differences between them, the fate of FA is qualitatively similar among species (IARC, 2006).

The major route of exposure to FA is inhalation. At room and body temperatures, FA hydrates rapidly and it is in equilibrium with its hydrated form methylene glycol (or methanediol) (NTP, 2011). Owing to its reactivity and solubility, FA is readily absorbed from the respiratory tract following inhalation. In obligate nose breathers, such as rodents, absorption and deposition occur primarily in the nasal passages. In humans, which are oronasal breathers, it is likely to occur primarily in the nasal passages and oral mucosa, but also in the trachea and bronchus (Kimbell et al., 2001). Species-specific differences of FA uptake and consequent associated effects (such as cell proliferation or reaction with other macromolecules) are determined by nasal airflow (nasal anatomy, ventilation) and breathing patterns rather than metabolism of the parent compound (Herausgegeben et al., 2006).

After oral exposure, FA is rapidly absorbed from the gastrointestinal tracts. In contrast, dermal uptake following application appears to be very slight, however it is able to penetrate skin (IARC, 2006).

FA is present at low levels in most living organisms. Physiological amounts of FA are endogenously formed from serine, glycine, methionine and choline, and also by demethylation of *N*-, *O*- and *S*-methyl compounds (IARC, 2006).

Nevertheless, regardless of its origin, endogenous or exogenous, FA has the same metabolic pathway in the organism (Figure 1) (ATSDR, 1999; IARC, 2006).

After absorption, FA is rapidly metabolised by oxidation to formate. The detoxification process takes place in all cells and can involve different enzyme systems. The primary and most important enzyme for the metabolic inactivation of FA is the glutathione-dependent formaldehyde-dehydrogenase (FDH) (also known as alcohol dehydrogenase class 3[1]; Koivusalo et al., 1989). FDH is highly specific for the FA-glutathione adduct, *S*-hydroxymethylglutathione, formed spontaneously in the cytosol from FA and glutathione conjugation. Formation of FA-glutathione adduct efficiently counteracts the presence of free FA, a reaction that is determined by the fact that glutathione is an abundant molecule in the cell. FDH oxidises *S*-hydroxymethylglutathione to *S*-formylglutathione, in the presence of NAD. This intermediate is then metabolised by *S*-formylglutathione hydrolase to yield formate and reduced glutathione (IPCS, 1989).

The activities of FDH are two to three orders of magnitude lower than those of *S*-formylglutathione hydrolase and thus the FDH-catalysed step is rate-limiting (IARC, 2006).

A secondary pathway involves the two aldehyde dehydrogenases (ALDH), ALDH1A1 (in cytosol) and ALDH2 (in mitochondria). Both enzymes have affinity for free FA but their *Km* values are higher than the *Km* displayed by FDH for *S*-hydroxymethylglutathione (Just et al., 2011). Therefore, FDH is probably the predominant enzyme responsible for the oxidation of FA at physiologically relevant concentrations, while ALDH enzymes contribute increasingly when the concentrations of FA increase (Just et al., 2011). Furthermore, some studies showed that FDH activity do not increase in response to FA exposure (Casanova-Schmitz et al., 1984; Øvrebø et al., 2002) thus no increase in metabolism occurs and non-metabolised FA is probably bound to macromolecules.

[1] The systematic name for the enzyme is formaldehyde: NAD+ oxidoreductase (IUBMB code EC 1.2.1.46). Other names for this enzyme are formaldehyde dehydrogenase, NAD-linked formaldehyde dehydrogenase and NAD-dependent formaldehyde dehydrogenase.

Other alternative metabolic pathways include, FA oxidation to formate by catalase, but only in the presence of hydrogen peroxide (H_2O_2); and FA reduction to methanol by a cytosolic alcohol dehydrogenase class 1 (ADH1) (IARC, 2006) (Figure 1).

Being normal components of intermediary metabolism, neither FA nor formate are stored to any significant extent in any tissue. Formate is either excreted in the urine, as formic acid, or oxidised to carbon dioxide and exhaled.

Moreover, both FA and formate can be further incorporated into other cellular molecules, entering the 1 carbon intermediary metabolic pool, which involves the folic acid metabolic pathway for synthesis of nucleic acids, amino acids, and macromolecules (ATDSR, 1999).

The kinetics of FA metabolism and the effects on its rate of various enzyme inhibitors have been studied (IARC, 2006). In isolated rat hepatocytes, depletion of glutathione or inhibition of the metabolic enzymes, FDH, ALDH2 or ADH1 were found to decrease FA metabolism and to increased cytotoxicity in a dose-dependent manner, through reactive oxygen species (ROS) production (Teng et al., 2001). Recent studies have identified ALDH2 as the primary enzyme responsible for the detoxication of aldehydes in haematopoietic stem cells, the lack of this enzyme results in hypersensitivity of haematopoietic stem cells to exogenous aldehydes, by accumulating DNA damage (Langevin et al., 2011; Garaycoechea et al., 2012). Thus, it is reasonable to assume that individuals with lower levels of glutathione as well as those with deficiencies in any of the above enzymes (by altered genes or polymorphisms) will be more susceptible to FA toxicity.

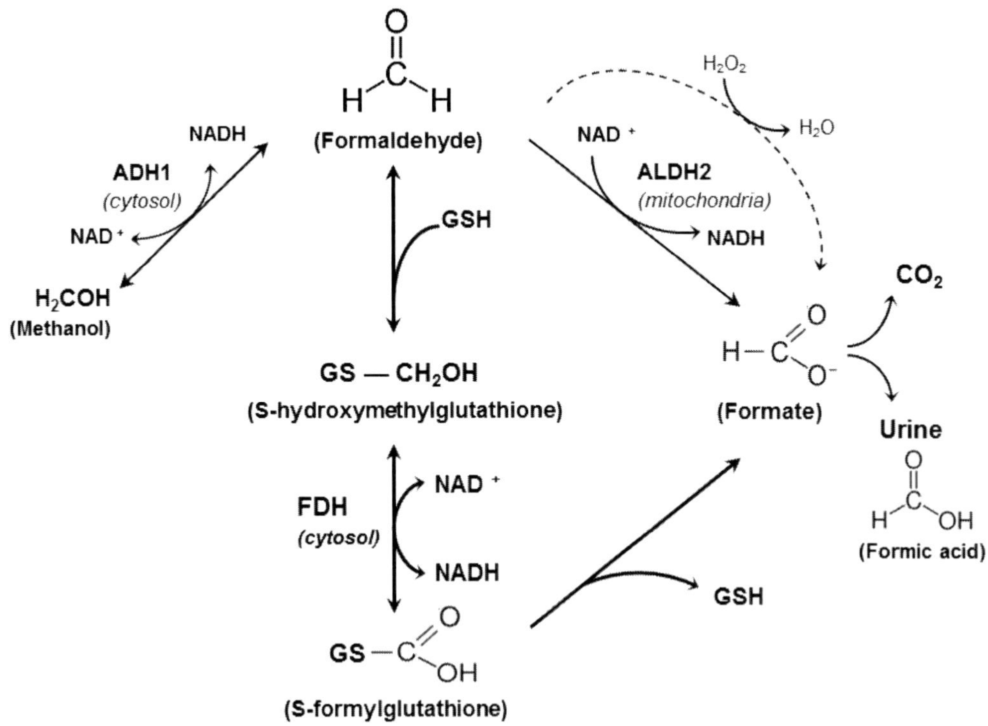

Figure 1. Fate and metabolism of formaldehyde (*adapted from Teng* et al. *(2001) and IARC (2006)*).

4. HEALTH EFFECTS

4.1. Sensory Irritation and Related Symptoms

FA has a pungent and suffocating odour and at certain concentrations is a strong irritating gas (ATDSR, 2010).

The respiratory tract, especially the upper respiratory tract, is a critical target of the irritant action of inhaled-FA, as shown by human acute controlled exposure studies, by studies on humans exposed acutely or repeatedly under occupational or residential conditions, and by studies of animals exposed by inhalation for acute, intermediate, and chronic durations (IARC, 2006; ATDSR, 2010).

Most humans detect and/or recognise FA odour at concentrations below 1 ppm (ATDSR, 2010), but the odour threshold has been reported to be between 0.05 and 0.18 ppm (Arts et al., 2006). The most common exposure effects to FA vapours comprise sensory irritation of the eyes, nose and throat, with eye irritation generally accepted as the most sensitive endpoint.

Odour perception and the threshold for sensory irritation are sometimes not clearly differentiated, it generally exist on a concentration gradient, but may overlap (Gaffney and Paustenbach, 2007). In FA case, however, the progression of responses is clear, with human studies showing that for most individuals, odour perception generally precedes sensory irritation (Golden, 2011). Importantly, odour perception and sensory irritation are different and distinct physiological processes. Upon substance inhalation odour is the sensation of smell carried by the olfactory nerve (first cranial nerve), while sensory irritation involves stimulation of the trigeminal nerve endings, and near triggering concentrations is generally considered to be a physiological and not a toxic response (Gaffney and Paustenbach, 2007; Golden, 2011).

Sensory irritation encompasses a graded series of responses related to the concentration of FA in the air and is generally categorised as ranging from slight to moderate (mucosa irritation) to severe (e.g., difficulty in breathing). The response also depends on the length of exposure and individual sensitivity.

The most common symptoms, irritation of the eyes, nose, and throat, along with increased tearing, occurs at FA-concentrations of about 0.3 - 3 ppm in the air.

Other symptoms described by individuals exposed to FA include sneezing, coughing, sinus, sore throat and nausea (Lang et al., 2008). In fact, clinical symptoms of eye soreness (68%), lacrimation (60%), eye fatigue (45%), rhinorrhea (38%), and throat irritation (43%) were reported in a survey involving 143 medical students exposed to a mean FA concentration of 2.4 ppm, during a course (15 hours/week for 2 months) (Takahashi et al., 2007). Similar symptoms were also described by students but at a lower airborne FA concentration, 0.72 ppm (6-8 hours/day for 3 months) (Wei et al., 2007).

Neurological effects due to FA exposure were also reported by a few studies namely, decrease performance in short-term memory tests and in the ability to concentrate (Bach et al., 1990; ATDSR, 2010). Indeed, in animal studies impaired learning/memory and altered motor activity were reported following acute- and intermediate-duration exposure to FA (ATDSR, 2010).

Changes in human pulmonary function, detected by spirometry tests (FEV, FVC), are assumed to occur at 0.6 ppm (ATDSR, 2010), but probably as protective response of the organism to the exposure, since the exposure to sensory irritants of the upper respiratory tract often cause slower breathing or even breath holding (Gaffney and Paustenbach, 2007). Volunteers (healthy and asthmatic) in acute controlled exposure studies showed no significant changes in pulmonary function after exposure to FA. In chronic exposures, some authors reported a decrease in pulmonary function of workers exposed to FA (Akbar-Khanzadeh et al., 1994; Fransman et al., 2003; Tang et al., 2009), while others found no exposure-related differences (Arts et al., 2006; ATDSR, 2010).

Increased numbers of eosinophils and protein exudation following FA exposure (0.4 ppm) were found in nasal lavage studies confirming the irritative effect of FA; a non-specific pro-inflammatory effect of FA was suggested by the authors (Pazdrak et al., 1993).

Nasal epithelial lesions have been observed consistently across a few occupational studies (ATDSR, 2010). In addition, clinical findings of upper respiratory tract inflammation were reported in 41% of workers exposed to FA (0.71 ppm) namely, hypertrophy or atrophy of the upper respiratory mucous membranes, chronic pharyngitis, rhinitis, rhinosinusitis, and rhinopharyngitis (Lyapina et al., 2004).

4.2. Sensitisation

FA (vapour) is considered an etiologic factor in occupational asthma, but although it may cause asthma in some individuals, this occurs rarely (Ezratty et al., 2007) and it cannot be regarded as a potent asthmogenic agent (Kim et al., 2001).

In fact, over the years, the association between FA-exposure and asthma (or exacerbation of) has been controversial, with investigations providing limited evidence of a causal relationship.

Nevertheless, a few case-reports of asthma associated to occupational exposure to FA were confirmed by bronchial provocation tests (Kim et al., 2001; IARC, 2006; Tang et al., 2009). The mechanism of FA-induced asthma remains unclear (Kim et al., 2001). However, it seems to be related to a delayed-type hypersensitivity reaction, since the subjects showed a latent period of symptomless reaction after exposure (IARC, 2006).

A review on controlled chamber studies involving asthmatic volunteers found that acute exposures to FA up to 3 ppm are unlikely to provoke asthma in an unsensitised asthmatic (Liteplo and Meek, 2003), yet high levels of FA may probably cause asthmatic reactions by an irritant mechanism (IARC, 2006).

Interestingly, a non-occupational study reported a significant relationship between asthma in children and residential indoor concentration of FA, at low levels (above 0.05 ppm) (Rumchev et al., 2002). This finding was recently explored by a meta-analysis of seven peer reviewed studies which confirmed the positive association between indoor inhalation of FA and induction/exacerbation of asthma in children (McGwin et al., 2010). Some authors however stated that these reports were confounded by potential co-exposures and other factors (Arts et al., 2006; Golden, 2011). Among adults, a few studies have reported a positive association between indoor levels of FA and asthma while others have not (McGwin et al., 2010). The effects of FA-indoor levels appear to be less evident in adults than in children (especially in asthmatics) living in the same household (Krzyzanowski et al., 1990).

The question whether FA vapour exposure is a risk factor for occupational or childhood asthma (or its exacerbation), or not, or may act in concert with other identified or as yet unidentified factors, still remains unanswered (Golden, 2011). Further data from other studies suggested a possible association between exposure to FA and allergic sensitisation to common aeroallergens in children and in asthmatic subjects (Garrett et al., 1999; Casset et al., 2006) however a recent study found no significant deleterious effect on airway allergen responsiveness of subjects with intermittent asthma (Ezratty et al., 2007).

FA (formalin solution) is a known skin sensitizer and one of the common causes of allergic contact dermatitis (IARC, 2006). Skin sensitivity to FA has been associated with many situations of dermal exposure, including contact with formalin, FA-containing resins, FA-treated fabrics and FA containing household products (Kim et al., 2001). Furthermore, FA has been widely reported to cause dermal allergic reactions in occupationally exposed workers, namely, laboratory workers, hairdressers, textile workers and industry workers (Kim et al., 2001; Aalto-Korte, 2008).

4.3. Genotoxicity

In the last two decades, a large number of studies were published on the genotoxicity of FA in a variety of biological systems and endpoints. The vast majority demonstrate FA genotoxicity in a multiple of experimental *in vitro* and *in vivo* models and in exposed humans.

In fact, FA has given positive results for almost all genetic endpoints evaluated in bacteria, non-mammalian eukaryotes (yeast, fungi, plants, insects, nematodes), *in vitro* studies with proliferating mammalian cell lines, and *in vivo* animal studies (IARC, 2006; NTP, 2011).

The evidence of FA-induced mutations *in vitro* experimental systems is consistent and generally associated with clastogenic effects and direct DNA mutation.

FA showed mutagenic potential in *Salmonella typhimurium* and *Escherichia coli*, with or without metabolic activation, including forward and reverse mutations (Schmid et al., 1986; IARC, 2006) and microsatellite instability (Wang et al., 2007).

Studies in non-mammalian eukaryotes and plants were also positive for forward and reverse mutations, lethal mutations and DNA single-strand breaks (IARC, 2006). For example, yeast strains deficient in repair pathways showed to be more susceptible to the mutagenic effects of FA (Magaña-Schwencke and Ekert, 1978).

Similarly, *in vitro* treatment of mammalian cells with different concentrations of FA induced several genotoxic endpoints including, deletions, point mutations, insertions, cell transformation, cytogenetic alterations (chromosomal aberrations, sister chromatid exchanges and micronucleus), DNA-protein crosslinks, DNA single-strand breaks and DNA- adducts (IARC, 2006).

In fact, chromosomal aberrations (CAs), sister chromatid exchanges (SCEs) and micronucleus (MN) were all increased *in vitro* in several human and rodent primary cells and cell lines (IARC, 2006, 2012). Schmid and Speit (2007) observed elevated MN formation in human lymphocytes isolated from whole blood cultures after *in vitro* exposure to FA (at 44h). Further evidence of FA-induced MN was found in DNA-repair-deficient cell lines (Ridpath et al., 2007), indicating the importance of certain repair pathways in the prevention of DNA lesions caused by FA exposure. A549 human lung cells and V79 Chinese hamster cells showed increased frequency of SCEs after treatment with different concentrations of FA. One hour after the addition of the agent to the A549 cells, the culture medium still retained the capacity to produce SCEs in non-exposed V79 cells, suggesting that genotoxicity persists despite the high reactivity of FA with macromolecules in the culture medium (Neuss and Speit, 2008; IARC, 2012).

Other *in vitro* studies described the formation of DNA-protein crosslinks in mammalian cells treated with FA (Conaway et al., 1996). Some of the proteins enrolled in these lesions were identified in cellular functions, namely transcription, gene regulation, DNA replication, and DNA repair (Qiu and Wang 2009).

It has also been reported that FA is able to induce hydroxymethyl adducts in mammalian DNA. For instance, DNA adducts were formed in an experimental reaction with FA and deoxyribonucleosides (Cheng et al., 2008a), DNA from human placenta (Zhong and Hee 2004) and DNA from human nasal epithelial cells (Speit et al., 2008). In addition, Lu et al. (2009) demonstrated the formation *in vitro* of an adduct between DNA and *S*-hydroxymethylglutathione, a key intermediate in cellular FA detoxication (please see 2.4).

Laboratory animals inhaling FA have generally shown more genotoxic effects in the nasal tissues. In rats FA vapour inhalation resulted in CAs in pulmonary lavage cells (Dallas et al., 1992), MN induction in the gastrointestinal tract (Migliore et al., 1989), DNA-protein crosslinks in the nasal mucosa and fetal liver (IARC, 2006), DNA strand breaks in lymphocytes and liver (Im et al., 2006; Wang and Liu, 2006) and dominant lethal mutations (Odeigah PG, 1997). However, no significant increase in CAs, MN or SCE frequencies were found in peripheral blood of rats (IARC, 2006, 2012).

A number of human biomonitoring studies have examined the genotoxic effect of FA in occupationally exposed populations, with both positive and negative outcomes. A wide variety of genotoxic endpoints has been investigated, particularly cytogenetic alterations, DNA damage and more recently DNA adducts. A fundamental aspect in the assessment of genotoxic effects of FA in exposed humans is whether genotoxicity is limited to first contact tissues or not. Therefore, genotoxic evaluations had focused on local-site tissues such as nasal and buccal mucosa cells or in circulating blood lymphocytes, nevertheless, there are few studies published that combined these two type of tissues (Ye et al., 2005; Viegas et al., 2010; Costa et al., 2013).

Several studies have reported a significant increased frequency of MN (buccal, nasal and lymphocytes) (Suruda et al., 1993; He et al., 1998; Orsière et al., 2006), SCEs (He et al., 1998; Shaham et al., 1997, 2002), CAs (He et al., 1998, Jakab et al., 2010), DNA-protein crosslinks (Shaham et al., 1997, 2003) and DNA damage (by Comet assay) (Yu et al., 2005; Costa et al., 2008) in FA-exposed workers, namely, industry workers, embalmers and pathology and anatomy workers.

A few studies have identified and quantified FA adducts in occupational exposed individuals and in smokers (FA is one of the main components of tobacco). Pala et al. (2008) reported a significant increase in FA- serum albumin conjugates in a small group of FA-exposed laboratory workers compared to referents. Further, a recent study by Bono et al. (2010) found a significant increase on the frequency of leukocyte malondialdehyde-deoxyguanosine adducts (M1-dG), a biomarker of oxidative stress and lipid peroxidation in a group of FA-exposed pathologists compared to controls. In smokers ($\geq$ 10 cigarettes/day), Wang et al. (2009) has identified a specific FA-DNA adduct. The N^6-hydroxymethyldeoxyadenosine (N^6-HOMe-dAdo) adduct was detected in 29 of the 32 smokers and in 7 of the 30 non-smokers, the difference was statistically significant. Furthermore, besides the tobacco smoke, authors proposed other minor sources of FA, namely as a by-product of another tobacco-specific compound or as result of lipid peroxidation or inflammation.

Also of note, a set of 89 miRNAs were significantly dysregulated in human lung cells treated with FA. Mapping of changed miRNAs and predicted mRNA targets revealed an association between FA exposure and inflammatory response pathways (Rager et al., 2011).

The formation of FA-DNA adducts in leukocytes of FA exposed subjects demonstrate that inhaled FA is able to enter systemic circulation and produce genotoxic effects in circulating cells, which may ultimately lead to carcinogenesis in distal sites.

Furthermore, it is believed that FA-genotoxicity and cytotoxicity have a key role in FA-related human carcinogenesis (NTP, 2011).

Genotoxic endpoints analysis, either cytogenetic alterations or DNA adducts, are of great interest in risk assessment of occupational carcinogens because they precede by a long time the potential health effects, thus offering a greater potential for preventive intervention (Mayeaux R., 2004).

4.4. Carcinogenicity

In 2004, the International Agency for Research on Cancer, IARC, reclassified FA from Group 2A, *probably carcinogenic to humans*, to Group 1, *carcinogenic to humans* (IARC, 2006). Recently, the US National Toxicology Programme (NTP) has revised FA carcinogenic status and after a rigorous scientific review, FA was listed in the NTP 12th Report on Carcinogens as *known to be a human carcinogen* (NTP, 2011). Both reclassifications are based on sufficient epidemiological evidence from human studies, experimental evidence in animals and supporting studies on mechanisms of carcinogenesis.

4.4.1. Epidemiological Studies

Numerous epidemiological studies have evaluated the relationship between cancer risk and exposure to FA in occupational settings, with focus on industry workers, embalmers and health professional groups namely physicians, pathologists and anatomists. Epidemiological studies have demonstrated a causal relationship between exposure to FA and cancer in humans (NTP, 2011). Causality is indicated by consistent findings of increased risks of nasopharyngeal cancer and leukaemia, among individuals with higher measures (exposure level and duration) of exposure to FA, which cannot be explained by chance, bias or unrecognized confounding effects (NTP, 2011; IARC, 2012). In addition, a positive association was observed between exposure to FA and sinonasal cancer (IARC, 2012). A number of studies have also found associations between FA-exposure and cancer at other sites, including the oral cavity, oro- and hypopharynx, pancreas, larynx, lung and brain, however, the overall evidence does not support a causal role for FA in these cancers (NTP, 2011; IARC, 2012).

Some of the more important studies implicated in FA carcinogenicity in humans are briefly described below.

Nasopharyngeal Cancer

The nasopharynx comprises the upper part of the throat that lies behind the nose. Studies have shown that all three histological types of nasopharyngeal cancer (undifferentiated carcinoma and keratinising and non-squamous cell carcinoma) start from the epithelial cells of the nasopharynx. The nasopharyngeal cancer is a rare cancer representing about 0.7% of the global cancer burden (Jemal et al., 2011).

The sufficient epidemiological evidence that FA causes nasopharyngeal cancer comes from five case–control studies and from the largest and most informative cohort study available to date involving workers from ten US industries (25 619 subjects), conducted by the National Cancer Institute (NCI).

In the NCI cohort study, a statistically significant excess of deaths from nasopharyngeal cancer was found in subjects occupationally exposed to FA in comparison with the general population (Hauptmann et al., 2004). Furthermore, a highly significant exposure–response correlation was observed between cancer mortality rate and peak (ppm) and cumulative exposures (ppm-years). Controlling for co-exposure to other potential occupational carcinogens and bias did not alter the exposure-response relationship. All cases were in the highest level of exposure category, most of which occurred in one plant, Plant 1.

A separate case-control study nested in NCI cohort was conducted to investigate if the excess mortality from nasopharyngeal cancer found in Plant 1 were related to other factors, namely, occupational history outside the FA industry. Marsh et al. (2007) found that silversmithing, prior to or after employment in Plant 1, was associated to nasopharyngeal cancer risk and concluded that the observed excess of nasopharyngeal cancers may be due to other occupational exposures rather than FA exposure, e.g silversmithing. However, several authors observed that while there was some indication of influence by activities as a silversmith, confounding alone did not explain the relatively high number of deaths from nasopharyngeal cancer found in this plant (Nielson et al., 2010; IARC, 2012). Furthermore, in Hauptmann et al. (2004) risks related to duration, cumulative, and average intensity of FA exposure did not decrease when silversmithing or other metal work were added as adjustment variables, indicating that silversmithing did not confound the association between FA and nasopharyngeal cancer.

Recently, the NCI cohort study was extended by 10 years (through 2004) the results from this update continue to indicate a link between FA occupational exposure and nasopharyngeal cancer, consistent with previous findings (Beane et al., 2013).

Increase incidences of deaths from nasopharyngeal cancer were also found in two other cohort studies: in a Danish study of proportionate cancer incidence, among subjects working in companies that manufactured or used FA (Hansen and Olsen, 1995) and in a proportionate mortality analysis of the largest US cohort of embalmers (Hayes et al., 1990).

Conversely, no excesses of nasopharyngeal cancer mortality was detect in other large cohort studies, two among industrial workers (Coggon et al., 2003; Pinkerton et al., 2004) and one among embalmers and funeral directors (Walrath and Fraumeni 1983). However, in these studies the number of cases were fewer than expected and the statistical power to detect an effect on nasopharyngeal cancer (rare cancer) was low (IARC, 2006).

Five of seven case-control studies showed consistent findings of increased risk of nasopharyngeal cancer among individuals with the highest or overall exposure to FA, but not all were statistically significant (Vaughan et al., 1986, 2000, Roush et al., 1987, West et al., 1993, Hildesheim et al., 2001).

The studies by Vaughan et al. (2000) (multi-center) and Hildesheim et al. (2001) are regarded as the most informative because of their size, their exposure assessment and more importantly the evaluation of potential confounders (IARC, 2012). In both case-control studies, risks of nasopharyngeal cancers increased with exposure duration (years) and cumulative (ppm-years) exposure to FA.

Furthermore, Vaughan et al. (2000) stratified the analysis by histological subtype of nasopharyngeal cancer and used several different measures of exposure to evaluate risk. A statistically significant exposure-response trend was found for squamous and unspecified epithelial carcinomas. The risk of nasopharyngeal cancer, all types, was also increased, but only for workers whose exposure duration were the longest, and not necessarily the highest.

On the other hand, Hildesheim et al. (2001) found some evidence of increasing risk of nasopharyngeal cancer with duration of exposure, but the observed trend did not reach statistical significance. However, a strong effect was observed for Epstein-Barr seropositive (EBV) individuals. In fact, a significant increase in risk of nasopharyngeal cancer was found among those ever-exposed to FA and EBV positive. It has been suggested that FA may have an indirect influence in EBV reactivation through deregulation of nitrosothiol homeostasis and in addition may also interact with the virus to promote epithelial cell transformation (Thompson and Grafström, 2009).

Overall, in the five case-control studies higher risks were found among FA-exposed individuals in the high-exposure groups (Vaughan et al., 1986, Roush et al., 1987) or with more years since first exposure (West et al., 1993). Some studies also reported that the risk remained elevated after consideration of/or adjustment for tobacco smoking (West et al., 1993; Vaughan et al., 2000; Hildesheim et al., 2001) or exposure to wood dust (Vaughan et al., 2000; Hildesheim et al., 2001).

However, other case-control studies did not find an association between FA exposure and nasopharyngeal cancer (Armstrong et al., 2000, Hauptmann et al., 2009).

A few meta-analyses with some of the above studies, cohort and/or case-controls, are available in the literature, but with contrasting results. Bosetti et al. (2008) did not found a significant increase in nasopharyngeal cancer risk in three cohort mortality studies. On the other hand, Bachand et al. (2010) found a near significant risk in a meta-analysis of seven case-control studies, while for six cohort studies no significant increase was detected (after exclusion of Plant 1 data of NCI cohort).

Lymphohematopoietic Cancer

An excess of mortality from leukaemia or combined lymphohematopoietic cancer (cancer of the lymphatic and blood-forming systems) has been consistently observed in epidemiological studies involving embalmers, funeral directors and health professionals, namely, pathologists and anatomists (IARC, 2006, 2012).

Several of these studies, however, only examined cancer outcomes by occupation and occupational characteristics (e.g., duration of employment) without exposure assessment. A recent nested case-control study by Hauptmann et al. (2009) examined exposure-response relationships for lymphohematopoietic malignancies among embalmers and funeral directors by metrics of exposure (average, cumulative, peak, and duration of embalming) based on lifetime work histories, practices and predicted FA-exposure levels. Mortality from myeloid leukaemia was significantly elevated among ever-embalmers and significant trends were observed with duration and peak exposure to FA.

Three of the largest cohorts in industrial settings were updated, and two found evidence of an excess of deaths from leukaemia and lymphohematopoietic cancer associated with FA exposure, namely in the US NCI industrial cohort (Hauptmann et al., 2003; Beane et al.,

2009) and in the US cohort of garment workers (Pinkerton et al., 2004). Conversely, no excess of mortality for leukaemia was found among workers exposed to FA in a British industrial cohort study (Coggon et al., 2003).

Overall, the above studies found positive exposure-response relationships for combined lymphohematopoietic cancer or specific subtypes, but the strongest association observed was for myeloid leukaemia (NTP, 2011). For instance, in Meyers et al. (2013) although myeloid leukaemia mortality was elevated in garment workers compared to the US population, the overall leukaemia mortality was not.

The large embalmers nested case-control study by Hauptmann et al. (2009) and the cohort studies updates of NCI industrial workers (Hauptmann et al., 2003; Beane et al., 2009) are considered to be the most informative studies for myeloid risk evaluation related to FA exposure (NTP, 2011). In these studies, statistically significant exposure-response relationships were observed with peak, average exposure and duration of exposure to FA. Controlling for co-exposure to other potential occupational carcinogens did not alter the findings.

In Beane's et al., (2009) follow-up study the overall risk of myeloid leukaemia although elevated was declined compared to previous reports. This finding is consistent with largest risks occurring closer in time to relevant exposure (Beane et al., 2009) and resembles the wavelike pattern of the relative risks over time seen for known leukaemogenic agents (e.g., benzene) (Beane et al., 2009; NTP, 2011).

Few case control studies have assessed the relationship between FA-exposure and risk for leukaemia (Linos et al., 1990; Partanen et al., 1993; Blair et al., 2001). No clear association was found in any of the studies, probably due to the small number of exposed cases and limited statistical power to analyse exposure-response associations (Zhang et al., 2009).

Three meta-analysis were recently published with contrasting results. The meta-analysis by Zhang et al. (2009), which included 15 cohort and case control studies found a significantly elevated risk of leukaemia, specially, myeloid leukaemia. In opposition, a meta-analysis by Bachand et al. (2010) did not find a significantly elevated risk for leukaemia or subtypes associated to FA-exposure in the studies selected for analysis.

Bosetti et al. (2008) found an elevated risk of leukaemia across studies of professional groups (pathologists, anatomists and embalmers) but not across studies of industrial workers. This finding is consistent with observations by Hauptmann et al. (2009) that embalmers have longer duration of FA-exposure, higher cumulative exposure and are more likely to be exposed to peak exposure levels greater than 4 ppm than industrial workers.

Sinonasal Cancer

Sinonasal cancer is a rare disease with an annual incidence of about 1 per 100 000 and it comprise cancers of the nasal cavity and paranasal sinus (Luce et al., 2002).

A number of case–control studies have shown a positive association between sinonasal cancers and occupational exposure to FA (Olsen et al., 1984, Olsen and Asnaes 1986, Roush et al., 1987, Luce et al., 1993). In most studies, the significant increase of sinonasal cancer risk was associated with FA exposure–response patterns, even among workers with low or no exposure to wood dust or after adjusting for exposure to wood (Olsen et al., 1984, Olsen and Asnaes 1986, Luce et al., 1993, 2002).

Elevated risks were found for the two major types of sinonasal cancer types, adenocarcinoma and squamous-cell carcinoma, but in some studies, the association was stronger for the adenocarcinoma (Luce et al., 1993, 2000; NTP, 2011).

The increased risk of adenocarcinoma among FA-exposed workers was confirmed in a pooled analysis study, which collected and re-analysed data from 12 case-control studies (Luce et al., 2002), furthermore the increased risk was followed by an increased cumulative exposure (ppm-years).

Other case-control studies did not find an association between FA exposure and sinonasal cancer (Vaughan et al., 1986; IARC, 2012) due probably to the small sample size (NTP, 2011).

Collins et al. (1997) meta-analysis found a significant association between FA-exposure and mortality from sinonasal cancers, nonetheless the author pointed out that the result was probably confounded by concomitant wood dust exposure detected in few of the studies included.

Mix results were found with regard to cohort studies. A statistically significant excess of sinonasal cancer was obtained in a Danish cohort study among workers exposed to FA (Hansen and Olsen, 1995) and a non-significant excess of deaths from sinonasal cancer was found in the NCI cohort, among industry workers exposed to FA (Hauptmann et al., 2004).

No increase in deaths from sinonasal cancers was found in other large cohort studies, (industry workers, Coggon et al., 2003; garment workers, Pinkerton et al., 2004) nor in a meta-analyses of eight industrial cohort studies (Bosetti et al., 2008).

It should be noted that in these cohort studies the statistical power to detect an association between FA exposure and sinonasal cancer (rare cancer) was limited, and even the larger cohort studies had insufficient numbers of exposed workers and expected deaths (e.g., approximately three in the NCI cohort) (NTP, 2011).

Nonetheless, based on the data available the NTP commission found that there was sufficient epidemiological evidence that FA occupational exposure cause sinonasal cancer in humans (NTP, 2011). On the other hand, the IARC (2006, 2012) working group concluded that the existing studies only offered limited positive evidence of a causal relationship between FA exposure and sinonasal cancer (IARC, 2006). The major limitations identified were the concomitant exposure to wood-dust, strongly associated with sinonasal cancer in some case-control studies, and the discordant results between the cohort and case control studies, which may be due to lack of statistical power or could indicate the confounding effect of wood dust (IARC, 2012).

4.4.2. Animal Studies

FA exposure caused tumours in two rodent species (rats and mice) at several different tissue sites and by two different routes of exposure (inhalation and ingestion) (NTP, 2011).

First evidence of the carcinogenic potential of FA came from long-inhalation studies with rodents (IPCS, 1989). These studies demonstrate that high concentrations of FA can cause irreversible damage to the nasal epithelium of rats and that in some cases rats exposed continually to these concentrations developed neoplasia (Merck and Speit, 1998).

In five studies the long-term inhalation exposure to FA caused nasal tumours, both benign (polypoid adenoma) and malignant (predominantly squamous-cell carcinoma but also adenocarcinoma and carcinoma) in two rats strains: male and female F344 rats (Monticello et al., 1996, Kamata et al.,1997) and in male Sprague-Dawley rats (Sellakumar et al., 1985;

IARC, 2006). Nasal lesions were also observed in male B6C3F1 mice (Kerns et al., 1983) and in male Wistar rats in a short-term exposure study (13 weeks) (Feron et al., 1988). Although the increased incidences of nasal tumours were not statistically significant in these last studies, they were considered exposure-related and biologically relevant because of the rarity of this type of tumour (NTP, 2011).

Studies on oral administration of FA in rats via drinking water showed to increase the incidence of fore-stomach tumours in male rats (IARC, 2006). Moreover, increased incidences of rare malignant gastrointestinal tumors were observed in female Sprague-Dawley rats exposed to FA *in utero* and throughout life via drinking water (Soffritti et al., 1989; IARC, 2006). In a similar study, the life-long ingestion of FA caused increased incidences of total malignant tumors in Sprague-Dawley rats including, testicular tumors in males, mammary-gland tumors in females and hemolymphoreticular tumors in both male and females of the high-exposure groups (Soffritti et al., 2002). IARC (2006) noted several limitations in these last two studies, namely the lack of historical control data, the misclassifications of lymphomas and leukaemia and discrepancies in tumour incidences.

No tumors were observed in skin by topical applications of FA solutions in mice (Iversen et al., 1986). With respect to other species, no tumors were found in monkeys exposed to FA by inhalation; probably because of the small number of animals exposed and short exposure duration (NTP, 2011).

4.4.3. Mechanisms of Carcinogenesis

The mechanisms by which FA causes cancer are not completely understood, but most likely involve multiple modes of action, such as DNA reactivity, gene mutation, epigenetic effects, glutathione depletion, oxidative stress, chromosomal breakage, cytotoxicity and aneuploidy (NTP, 2011).

Upper Airway Cancers

Mechanistic studies in humans and experimental animals support the evidence that FA causes nasopharyngeal (IARC, 2006; NTP, 2011) and sinonasal cancer (NTP, 2011). The existing data strongly indicate that genotoxicity plays a key role in the carcinogenesis of FA in nasal tissues and also that cellular replication (in response to FA-induced cytotoxicity) promotes the carcinogenic response of the upper airway tissues (Conaway et al., 1996; McGregor et al., 2006).

Indeed, several studies show that inhaled FA is capable of causing genetic damage in the nasal epithelium of both experimental animals (e.g., DNA- protein cross-links above 6 ppm) and humans (e.g., buccal and nasal MN) (Conaway et al., 1996; Viegas et al., 2010; Costa et al., 2013).

Moreover, computational fluid dynamic model studies that have predict and compare FA-local airflow in the nasal passages of experimental animals showed that airway deposition and cytotoxicity-induced by cellular proliferation are important variables in the carcinogenicity of FA to nasal cells (Kimbel et al., 2001).

In rats, regional FA flux was correlated with the anatomical distribution of FA-induced lesions (squamous metaplasia) and DNA-protein crosslinks (Georgieva et al., 2003). FA-cytotoxicity in the nasal epithelium may result in cellular proliferation and in the promotion of chemically induced spontaneous mutations which may contribute to a carcinogenesis process (Monticello and Morgan, 1997). Indeed, cell-proliferation rates were correlated with the site specificity of nasal tumors (Monticello et al., 1989; Monticello and Morgan, 1997). However, FA exposure also causes cytotoxicity and cellular proliferation at anatomical sites that are not thought to be the origin of the squamous-cell carcinoma, suggesting that factors other than cellular proliferation may play a role in FA-induced nasal cancers (NTP, 2011).

Lymphohematopoietic Cancer

As discussed above, several epidemiological studies found evidence of an association between exposure to FA and leukaemia. These findings have raised some discussion in the scientific community. Owing to its reactive nature it has been considered unlikely that inhaled FA is able to damage directly the stem cells in bone marrow as classical leukaemogens and cause leukaemia (Zhang et al., 2010a). In fact, some authors strongly stated that the fact of inhaled FA enter systemic circulation, reach bone marrow and cause leukaemia is biologically implausible and FA genotoxic and carcinogenic action are limited to local effects, in the area of first contact where is rapidly metabolised (Heck and Casanova, 2004; Franks, 2005; Speit, 2006). In addition, some studies have showed that the endogenous concentration of FA in the blood of humans, monkeys, and rats (2-3 μg/g) does not increase after inhalation (Heck and Casanova, 2004). Furthermore, the biological evidence of systemic (geno-) toxic effects of FA in both animal and human studies is still insufficient and conflicting, with both positive and negative outcomes (IARC, 2006, 2012).

However, it has been previously suggested that FA could potentially reach bone marrow directly in its hydrate methanediol form. Although endogenous FA is usually rapidly metabolised by GSH-dependent pathways, excess levels such as those encountered occupationally could in fact saturate metabolic capacity and lead to genotoxicity in DNA and chromosomes (Zhang et al., 2010b).

Additionally, there is consistent evidence for systemic distribution of FA in humans showed by increased levels of FA-albumin adducts in serum in occupational exposed subjects (Pala et al., 2008; Bono et al., 2010) and in smokers (Wang et al., 2009), as well as, numerous studies in humans and experimental animals that have demonstrated that inhaled FA can cause hematologic effects, genotoxicity in circulating lymphocytes (Costa et al., 2008, 2011), and cancer at distal sites (e.g., myeloid leukaemia).

Nevertheless, the mechanisms by which FA causes myeloid leukaemia in humans are not known. A recent work by Zhang et al. (2009, 2010b) suggested two possible pathways by which FA could act as a leukaemogen in humans. One, by damaging hematopoietic stem/progenitor cells circulating in peripheral blood and the other, by damaging primitive pluripotent stem cells present within the nasal turbinates and/or olfactory mucosa (Zhang et al., 2009). In both models, the damaged stem/progenitor cells would then travel to the bone marrow and become initiated leukemic stem cells, potentially developing into leukaemia following a latency period (Zhang et al., 2010b). Given the likely dynamics of stem cell turnover between the nasal/oral passages, blood and bone marrow, sufficient stem cells could be targeted through these two alternative models, focused around genotoxicity, to induce leukaemia, under conditions of chronic high FA exposure (such as occupational exposure) (Zhang et al., 2009).

Overall, the available evidence taken together does not indicate that such mechanisms are implausible (NTP, 2011; IARC, 2012). New *in vitro* data demonstrate FA's ability to induce genotoxicity in mouse erythropoietic cells and suppress human erythroid progenitor cells expansion supporting the potential of adverse effects of FA on hematopoietic stem/progenitor cells (Ji et al., 2014). Furthermore, MN formation in mouse erythrocytes occurred in a dose-dependent manner, suggesting that chromosomal damage may be one potential mechanism underlying FA induced leukemogenesis. Additionally, it was recently reported that a FA-detoxifying enzyme (ALDH2) is involved on aldehyde detoxication in haematopoietic stem cells (Langevin et al., 2011; Garaycoechea et al., 2012). Thus, the biological plausibility of FA-induced leukemogenesis cannot be excluded.

REFERENCES

Aalto-Korte, K., Kuuliala, O., Suuronen, K., and Alanko, K. 2008. Occupational contact allergy to formaldehyde and formaldehyde releasers. *Contact Dermatitis,* 59(5):280-9.

ACGIH (American Conference of Governmental Industrial Hygienists). 2013. *TLV's and BEI's – Based on the Documentation of the Threshold Limit Values for Chemical Substances and Physical Agents and Biological Exposure Indices.* ACGIH, 238 pp.

AFSSET (L'Agence française de sécurité sanitaire de l'environnement et du travail). 2008. Valeurs limites d'exposition en milieu professionnel. Évaluation des effets sur la santé et des methods de mesure des niveaux d'exposition sur le lieu de travail pour le formaldehyde. Maisons-Alfort, France, 62 pp.

Akbar-Khanzadeh, F., Vaquerano, M.U., Akbar-Khanzadeh, M., and Bisesi, M.S. 1994. Formaldehyde exposure, acute pulmonary response, and exposure control options in a gross anatomy laboratory. *American Journal of Industrial Medicine,* 26(1):61-75.

Akbar-Khanzadeh, F. and Pulido, E.V. 2003. Using Respirators and Goggles to Control Exposure to Air Pollutants in an Anatomy Laboratory. *American Journal Industrial Medicine,* 43:326-331.

Armstrong, R.W., Imrey, P.B., Lye, M.S., Armstrong, M.J., Yu, M.C. and Sani, S. 2000. Nasopharyngeal carcinoma in Malaysian Chinese: Occupational exposures to particles, formaldehyde and heat. *International Journal of Epidemiology,* 29, 991–998.

Arts, J.H., Rennen, M.A., and de Heer, C. 2006. Inhaled formaldehyde: evaluation of sensory irritation in relation to carcinogenicity. *Regulatory Toxicology and Pharmacology,* 44(2):144-60.

ATSDR (Agency for Toxic Substances and Disease Registry). 1999. *Toxicological Profile for formaldehyde. US Department of Health and Human Services,* Public Health Service, Atlanta GA, USA, 421 pp.

Bach, B., Pedersen, O.F., and Mohvave, L. 1990. *Human Performance During Experimental Formaldehyde Exposure, Environmental International* 16:105-113 [abstract].

Bachand, A.M., Mundt, K.A., Mundt, D.J., and Montgomery, R.R. 2010. Epidemiological studies of formaldehyde exposure and risk of leukemia and nasopharyngeal cancer: a meta-analysis. *Critical Reviews in Toxicology,* 40(2):85-100.

Beane Freeman, L.E., Blair, A., Lubin, J.H., Stewart, P.A., Hayes, R.B., Hoover, R.N., and Hauptmann, M. 2009. Mortality from lymphohematopoietic malignancies among workers in formaldehyde industries: the National Cancer Institute Cohort. *Journal of the National Cancer Institute,* 101(10):751-61.

Beane Freeman, L.E., Blair, A., Lubin, J.H., Stewart, P.A., Hayes, R.B., Hoover, R.N., and Hauptmann, M. 2013. Mortality from solid tumors among workers in formaldehyde industries: an update of the NCI cohort. *American Journal of Industrial Medicine,* 56(9):1015-26.

Binetti, R., Costamagna, F.M., and Marcello, I. 2006. Development of carcinogenicity classifications and evaluations: the case of formaldehyde. *Annali dell'Istituto superiore di sanità,* 42(2):132-43.

Blair, A., Zheng, T., Linos, A., Stewart, P.A., Zhang, Y.W., Cantor, K.P.2001. Occupation and leukemia: a population-based case-control study in Iowa and Minnesota. *American Journal of Industrial Medicine,* 40(1):3-14.

Bono, R., Romanazzi, V., Munnia, A., Piro, S., Allione, A., Ricceri, F., Guarrera, S., Pignata, C., Matullo, G., Wang, P., Giese, R.W., and Peluso, M. 2010. Malondialdehyde-deoxyguanosine adduct formation in workers of pathology wards: the role of air formaldehyde exposure. *Chemical Research in Toxicology,* 23(8):1342-8.

Bosetti, C., McLaughlin, J.K., Tarone, R.E., Pira, E., and La Vecchia, C. 2008. Formaldehyde and cancer risk: a quantitative review of cohort studies through 2006. *Annals of Oncology,* 19(1):29-43.

Casanova-Schmitz, M., David, R.M., and Heck, H.D. 1984. Oxidation of formaldehyde and acetaldehyde by NAD+-dependent dehydrogenases in rat nasal mucosal homogenates. *Biochemical Pharmacology,* 33(7):1137-42.

Casset, A., Marchand, C., Purohit, A., le Calve, S., Uring-Lambert, B., Donnay, C., Meyer, P., and de Blay, F. 2006. Inhaled formaldehyde exposure: effect on bronchial response to mite allergen in sensitized asthma patients. *Allergy.* 61(11):1344-50.

Cheng, G., Wang, M., Upadhyaya, P., Villalta, P.W., and Hecht, S.S. 2008. Formation of formaldehyde adducts in the reactions of DNA and deoxyribonucleosides with alpha-acetates of 4-(methylnitrosamino)-1-(3-pyridyl)-1-butanone (NNK), 4-(methylnitrosamino)-1-(3-pyridyl)-1-butanol (NNAL), and N- nitrosodimethylamine (NDMA*). Chemical Research in Toxicology,* 21(3): 746-751.

CDC (Centers for Disease Control and Prevention). 2008 (amended in 2010). Final Report on Formaldehyde Levels in FEMA-Supplied Travel Trailers, *Park Models, and Mobile Homes.* CDC, Atlanta, USA, 58 pp.

Coggon, D., Harris, E.C., Poole, J., and Palmer, K.T. 2003. Extended follow-up of a cohort of British chemical workers exposed to formaldehyde. *Journal of the National Cancer Institute,* 95(21):1608-1615.

Conaway, C.C., Whysner, J., Verna, L.K., and Williams, G.M. 1996. Formaldehyde mechanistic data and risk assessment: endogenous protection from DNA adduct formation. *Pharmacology and Therapeutics,* 71(1-2):29-55.

Costa, S., Coelho, P., Costa, C., Silva, S., Mayan, O., Santos, L.S, Gaspar, J. and Teixeira, J.P. 2008. Genotoxic damage in pathology anatomy laboratory workers exposed to formaldehyde. *Toxicology,* 252(1-3): 40-48.

Costa S., Brandão F., Coelho M., Costa C., Coelho P., Silva S., Porto B., and Teixeira J.P. 2013. *Micronucleus frequencies in lymphocytes and buccal cells in formaldehyde exposed workers.* WIT Transactions on Biomedicine and Health (Environmental Health Risk, EHR 2013), 16: 83-94.

Dallas, C.E., Scott, M.J., Ward, J.B. Jr, and Theiss, J.C. 1992. Cytogenetic analysis of pulmonary lavage and bone marrow cells of rats after repeated formaldehyde inhalation. *Journal of Applied Toxicology,* 12(3):199-203 [abstract].

Ezratty, V., Bonay, M., Neukirch, C., Orset-Guillossou, G., Dehoux, M., Koscielny, S., Cabanes, P.A., Lambrozo, J., and Aubier, M. 2007. Effect of formaldehyde on asthmatic response to inhaled allergen challenge. *Environmental Health Perspectives,* 115(2):210-4.

Feron, V.J., Bruyntjes, J.P., Woutersen, R.A., Immel, H.R., and Appelman, L.M. 1988. Nasal tumours in rats after short-term exposure to a cytotoxic concentration of formaldehyde. *Cancer Letters,* 39 (1):101-11.

FIOH (Finnish Institute of Occupational Health). 1998. *CAREX- international information system on occupational exposure to known and suspected carcinogens in 1990-93.* European Union, Helsinki. http://www.ttl.fi/en/chemical_safety/carex/countries/portugal/pages/default.aspx.

Franks, S.J. 2005. A mathematical model for the absorption and metabolism of formaldehyde vapour by humans. *Toxicology Applied Pharmacology,* 206(3):309-20.

Fransman, W., McLean, D., Douwes, J., Demers, P.A., Leung, V., and Pearce, N. Respiratory symptoms and occupational exposures in New Zealand plywood mill workers. *Annals of Occupational Hygiene,* 47(4):287–295.

Gaffney, S.H., and Paustenbach, D.J. A proposed approach for setting occupational exposure limits for sensory irritants based on chemosensory models. *Annals of Occupational Hygiene.* 51(4):345-56.

Garaycoechea, J.I., Crossan, G.P., Langevin, F., Daly, M., Arends, M.J., Patel, K.J. 2012. Genotoxic consequences of endogenous aldehydes on mouse haematopoietic stem cell function. *Nature,* 489:571-575.

Garrett, M.H., Hooper, M.A., Hooper, B.M., Rayment, P.R., and Abramson, M.J. 1999. Increased risk of allergy in children due to formaldehyde exposure in homes. *Allergy,* 54, 330–337.

Georgieva, A.V., Kimbell, J.S., and Schlosser, P.M. 2003. A distributed-parameter model for formaldehyde uptake and disposition in the rat nasal lining. *Inhalation Toxicology,* 15(14):1435-63.

Golden, R. 2011. Identifying an indoor air exposure limit for formaldehyde considering both irritation and cancer hazards. *Critical Reviews in Toxicology,* 41(8):672-721.

Hansen, J., and Olsen, J.H. 1995. Formaldehyde and cancer morbidity among male employees in Denmark. *Cancer Causes Control,* 6(4):354-60.

Hayes, R.B., Blair, A., Stewart, P.A., Herrick, R.F., and Mahar, H. 1990. Mortality of U.S. embalmers and funeral directors. *American Journal of Industrial Medicine,* 18(6):641-52.

Hauptmann, M., Lubin, J.H., Stewart, P.A., Hayes, R.B., and Blair, A. 2003. Mortality from lymphohematopoietic malignancies among workers in formaldehyde industries. *J Natl Cancer Inst* 95(21): 1615-1623.

Hauptmann, M., Lubin, J.H., Stewart, P.A., Hayes, R.B. and Blair, A. 2004. Mortality from solid cancers among workers in formaldehyde industries. *American Journal of Epidemiol.,* 159, 1117–1130.

Hauptmann, M., Stewart, P.A., Lubin, J.H., Beane Freeman, L.E., Hornung, R.W., Herrick, R.F., Hoover, R.N., Fraumeni, J.F. Jr, Blair, A., and Hayes, R.B. 2009. Mortality from lymphohematopoietic malignancies and brain cancer among embalmers exposed to formaldehyde. *Journal of the National Cancer Institute,* 101(24):1696-708.

Heck, Hd., and Casanova, M. 2004. The implausibility of leukemia induction by formaldehyde: a critical review of the biological evidence on distant-site toxicity. *Regulatory Toxicology and Pharmacology,* 40(2):92-106.

Herausgegeben, von A.S., Bernauer, U., Madle, S., Mielke, H., Herbst, U., Richter-Reichhelm, H.B., Appel, K.E., and Gundert-Remy U. 2006. Bundesinstitut für Risikobewertung (BfR) - *Assessment of the Carcinogenicity of Formaldehyde* [CAS No. 50-00-0]. Berlin, Germany, 156 pp.

Hildesheim, A., Dosemeci, M., Chan, C.C., Chen, C.J., Cheng, Y.J., Hsu, M.M., Chen, I.H., Mittl, B.F., Sun, B., Levine, P.H., Chen, J.Y., Brinton, L.A., and Yang, C.S. 2001. Occupational exposure to wood, formaldehyde, and solvents and risk of nasopharyngeal carcinoma. *Cancer Epidemiology, Biomarkers and Prevention,* 10(11):1145-53.

IARC (International Agency for Research on Cancer). 2006. IARC Monographs on the Evaluation of Carcinogenic Risks to Human, Volume 88: Formaldehyde, 2-Butoxyethanol and 1-tert-Butoxy-2-propanol. *World Health Organization,* Lyon, France, 39-325 pp.

IARC (International Agency for Research on Cancer). 2012. A Review of Human Carcinogens: Chemical Agents and Related Occupations, IARC Monographs on the Evaluation of Carcinogenic Risks to Humans, Part F of Volume 100. *World Health Organization,* Lyon, France, 401-435 pp.

IHS Chemical. 2012. Chemical Economics Handbook- Formaldehyde. http: //www.ihs.com/products/chemical/planning/ceh/formaldehyde.aspx [abstract]. (consulted on 17th of November, 2013)

Im H., Oh E., Mun J., Khim J.Y., Lee E., Kang H.S., Kim E., Kim H., Won N.H., Kim Y.H., Jung W.W., Sul D. 2006. Evaluation of toxicological monitoring markers using proteomic analysis in rats exposed to formaldehyde. *Journal of Proteome Research,* 5(6): 1354-66.

IPCS (International Programme on Chemical Safety).1989. Environmental Health Criteria 89: Formaldehyde. *World Health Organization,* Geneva, Switzerland, 219 pp.

IPCS (International Programme on Chemical Safety). 2002. Concise international chemical assessment document (CICAD); no. 40 - Formaldehyde. *World Health Organization,* Geneva, Switzerland, 74 pp.

IPQ (Instituto Português da Qualidade). 2007. NP-1796: *Segurança e Saúde do Trabalho, Valores limite de exposição profissional a agentes químicos.* IPQ, Portugal, 53 pp.

Jakab, M.G., Klupp, T., Besenyei, K., Biró, A., Major, J., and Tompa, A. 2010. Formaldehyde-induced chromosomal aberrations and apoptosis in peripheral blood lymphocytes of personnel working in pathology departments. *Mutation Research,* 698(1-2):11-7.

Jemal, A., Bray, F., Center, M.M., Ferlay, J., Ward, E., and Forman, D. 2011. Global cancer statistics. CA: *Cancer Journal Clinicians,* 61(2):69-90.

Ji, Z., Li, X., Fromowitz, M., Mutter-Rottmayer, E., Tung, J., Smith, M.T., Zhang, L. 2014. Formaldehyde induces micronuclei in mouse erythropoietic cells and suppresses the expansion of human erythroid progenitor cells. *Toxicology Letters,* 224(2):233-9.

Just, W., Zeller, J., Riegert, C., and Speit, G. 2011. Genetic polymorphisms in the formaldehyde dehydrogenase gene and their biological significance. *Toxicology Letters,* 207(2):121-7.

Kamata, E., Nakadate, M., Uchida, O., Ogawa, Y., Suzuki, S., Kaneko, T., Saito, M., and Kurokawa, Y. 1997. Results of a 28-month chronic inhalation toxicity study of formaldehyde in male Fisher-344 rats. *Journal of Toxicological Sciences,* 22(3):239-54.

Kerns, W.D., Pavkov, K.L., Donofrio, D.J., Gralla, E.J., and Swenberg, J.A. 1983. Carcinogenicity of formaldehyde in rats and mice after long-term inhalation exposure. *Cancer Research,* 43(9):4382-92.

Kim, C.W., Song, J.S., Ahn, Y.S., Park, S.H., Park, J.W., Noh, J.H., and Hong, C.S. 2001. Occupational asthma due to formaldehyde. *Yonsei Medical Jounal,* 42(4):440-5.

Kimbell, J.S., Overton, J.H, Subramaniam, R.P., Schlosser, P.M., Morgan, K.T., Conolly, R.B., and Miller, F.J. 2001. Dosimetry modeling of inhaled formaldehyde: binning nasal flux predictions for quantitative risk assessment. *Toxicological Sciences,* 64(1):111-21.

Koivusalo, M., Baumann, M., and Uotila, L. 1989. Evidence for the identity of glutathione-dependent formaldehyde dehydrogenase and class III alcohol dehydrogenase. *FEBS Letters,* 257(1):105-9

Krzyzanowski, M., Quackenboss, J.J., and Lebowitz, M.D. 1990. Chronic respiratory effects of indoor formaldehyde exposure. *Environmental Research,* 52(2):117-25

Lang, I., Bruckner, T., and Triebig, G. 2008. Formaldehyde and chemosensory irritation in humans: a controlled human exposure study. *Regulatory Toxicology and Pharmacology,* 50(1):23-36.

Langevin, F., Crossan, G.P., Rosado, I.V., Arends, M.J., and Patel, K.J. 2011. Fancd2 counteracts the toxic effects of naturally produced aldehydes in mice. *Nature,* 475(7354):53-8.

Linos, A., Blair, A., Cantor, K.P., Burmeister, L., VanLier, S., Gibson, R.W., Schuman, L., and Everett, G. 1990. Leukemia and non-Hodgkin's lymphoma among embalmers and funeral directors. *Journal of the National Cancer Institute,* 82(1):66.

Liteplo, R.G., and Meek M.E. 2003. Inhaled formaldehyde: exposure estimation, hazard characterization, and exposure-response analysis. *Journal of Toxicology and Environmental Health, Part B: Critical Reviews,* 6(1):85-114.

Liu, Y., Li, C.M., Lu, Z., Ding, S., Yang, X., Mo, J. 2006. Studies on formation and repair of formaldehyde-damaged DNA by detection of DNA-protein crosslinks and DNA breaks. *Frontiers in Bioscience,* 11:991-7.

Lu, K., Ye, W., Gold, A., Ball, L.M., and Swenberg, J.A. 2009. Formation of S-[1-(N2-deoxyguanosinyl)methyl]glutathione between glutathione and DNA induced by formaldehyde. *Journal of the American Chemical Society,* 131(10):3414-5.

Luce, D., Gérin, M., Leclerc, A., Morcet, J.F., Brugère, J., Goldberg, M. 1993. Sinonasal cancer and occupational exposure to formaldehyde and other substances. *International Journal of Cancer,* 53(2):224-31.

Luce, D., Leclerc, A., Bégin, D., Demers, P.A., Gérin, M., Orlowski, E., Kogevinas, M., Belli, S., Bugel, I., Bolm-Audorff, U., Brinton, L.A., Comba, P., Hardell, L., Hayes, R.B., Magnani, C., Merler, E., Preston-Martin, S., Vaughan, T.L., Zheng, W., and Boffetta, P. 2002. Sinonasal cancer and occupational exposures: a pooled analysis of 12 case-control studies. *Cancer Causes and Control,* 13(2):147-57

Lyapina, M., Zhelezova, G., Petrova, E., and Boev, M. 2004. Flow cytometric determination of neutrophil respiratory burst activity in workers exposed to formaldehyde. *International Archives of Occupational and Environmental Health,* 77(5): 335-340.

Magaña-Schwencke, N. and Ekert, B. 1978. Biochemical analysis of damage induced in yeast by formaldehyde. II. Induction of cross-links between DNA and protein. *Mutation Research,* (51):11–19 [abstract].

Marsh, G.M., Youk, A.O., Buchanich, J.M., Erdal, S., and Esmen, N.A. 2007. Work in the metal industry and nasopharyngeal cancer mortality among formaldehyde-exposed workers. *Regulatory Toxicology and Pharmacology,* 48: 308–319.

Mayeux R. 2004. Biomarkers: potential uses and limitations. *NeuroRx* 1(2):182-8.

McGregor, D., Bolt, H., Cogliano, V., and Richter-Reichhelm, H.B. 2006. Formaldehyde and glutaraldehyde and nasal cytotoxicity: case study within the context of the 2006 IPCS Human Framework for the Analysis of a cancer mode of action for humans. *Critical Reviews of Toxicology,* 36(10): 821-835.

McGwin, G., Lienert, J., and Kennedy, J.I. 2010. Formaldehyde exposure and asthma in children:a systematic review. *Environmental Health Perspectives,* 118(3):313-7.

Meyers, A.R., Pinkerton, L.E., and Hein, M.J. 2013. Cohort mortality study of garment industry workers exposed to formaldehyde: update and internal comparisons. *American Journal of Industrial Medicine,* 56(9):1027-39.

Migliore, L., Ventura, L., Barale, R., Loprieno, N., Castellino, S., and Pulci, R.1989. Micronuclei and nuclear anomalies induced in the gastro-intestinal epithelium of rats treated with formaldehyde. *Mutagenesis,* 4(5):327-34.

Monticello, T.M., Morgan, K.T., Everitt, J.I., and Popp, J.A. 1989. Effects of formaldehyde gas on the respiratory tract of rhesus monkeys. Pathology and cell proliferation. *American Journal Pathology,* 134(3):515-27.

Monticello, T.M., Swenberg, J.A., Gross, E.A., Leininger, J.R., Kimbell, J.S., Seilkop, S., Starr, T.B., Gibson, J.E., and Morgan KT. 1996. Correlation of regional and nonlinear formaldehyde-induced nasal cancer with proliferating populations of cells. *Cancer Research,* 56(5):1012-22.

Monticello, T.M., and Morgan, K.T. 1997. Chemically-induced nasal carcinogenesis and epithelial cell proliferation: a brief review. *Mutation Research,* 380 (1-2):33-41.

NTP (National Toxicology Program). 2011. *Report on Carcinogens, Twelfth Edition-Formaldehyde.* Research Triangle Park, NC: US Department of Health and Human Services, Public Health Service, National Toxicology Program, 195-205 pp.

Nielsen, G.D., and Wolkoff, P. 2010. Cancer effects of formaldehyde: a proposal for an indoor air guideline value. *Archives of Toxicology,* 84(6):423-46.

Odeigah, P.G. 1997. Sperm head abnormalities and dominant lethal effects of formaldehyde in albino rats. *Mutation Research,* 389(2-3):141-8.

Olsen, J.H., and Asnaes, S. 1986. Formaldehyde and the risk of squamous cell carcinoma of the sinonasal cavities. *British Journal of Industrial Medicine,* 43(11):769-74.

Olsen, N.J., Visell, S.M., Fan, J., and Kovacs, W.J. 1998. Androgens accelerate thymocyte apoptosis. *Endocrinology,* 139(2):748-52.

Orsière, T., Sari-Minodier, I., Iarmarcovai, G., and Botta, A. 2006. Genotoxic risk assessment of pathology and anatomy laboratory workers exposed to formaldehyde by use of personal air sampling and analysis of DNA damage in peripheral lymphocytes. *Mutation Research,* 605(1-2): 30-41.

OSHA (Occupational Safety and Health Standards). 2011. OSHA fact sheet-formaldehyde. OSHA, 2pp. https://www.osha.gov/OshDoc/ data_ General_Facts/formaldehyde-factsheet.pdf. (consulted on 17th of November, 2013).

Øvrebø, S., Haugen, A., and Skaug, V. 2002. Biotransformation of formaldehyde in cultured human bronchus. *Environmental Research,* 89(1):38-42.

Pala, M., Ugolini, D., Ceppi, M., Rizzo, F., Maiorana, L., Bolognesi, C., Schilirò, T., Gilli, G., Bigatti, P., Bono, R., and Vecchio, D. 2008. Occupational exposure to formaldehyde and biological monitoring of Research Institute workers. *Cancer Detection and Prevention,* 32(2):121-6.

Partanen, T., Kauppinen, T., Luukkonen, R., Hakulinen, T., Pukkala, E. 1993. Malignant lymphomas and leukemias, and exposures in the wood industry: an industry-based case-referent study. *International archives of occupational and environmental health* 64(8):593-6.

Pazdrak, K., Górski, P., Krakowiak, A., and Ruta, U. 1993. Changes in nasal lavage fluid due to formaldehyde inhalation. *International Archives of Occupational and Environmental Health* 64(7):515-9 [abstract].

Pierce, J.S., Abelmann, A., Spicer, L.J., Adams, R.E., Glynn, M.E., Neier, K., Finley, B.L., and Gaffney, S.H. 2011. Characterization of formaldehyde exposure resulting from the use of four professional hair straightening products. *Journal of Occupational and Environmental Hygiene.* 8(11):686-99.

Pilidis, G.A., Karakitsios, S.P., Kassomenos, P.A., Kazos, E.A., and Stalikas, C.D. 2008. Measurements of benzene and formaldehyde in a medium sized urban environment. Indoor/outdoor health risk implications on special population groups. *Environmental Monitoring Assesment,* 150(1-4):285-94.

Pinkerton, L.E., Hein, M.J., Stayner, L.T. 2004. Mortality among a cohort of garment workers exposed to formaldehyde: an update. *Occupational and Environmental Medicine,* 61(3): 193-200.

Qiu, H., and Wang, Y. 2009. Exploring DNA-binding proteins with in vivo chemical cross-linking and mass spectrometry. *Journal Proteome Research,* 8: 1983-1991.

Rager, J.E., Smeester, L., Jaspers, I., Sexton, K.G., and Fry, R.C. 2011. Epigenetic changes induced by air toxics: formaldehyde exposure alters miRNA expression profiles in human lung cells. *Environmental Health Perspectives,* 119(4):494-500.

Ridpath JR, Nakamura A, Tano K, Luke AM, Sonoda E, Arakawa H, Buerstedde JM, Gillespie DA, Sale JE, Yamazoe M, Bishop DK, Takata M, Takeda S, Watanabe M, Swenberg JA, Nakamura J. 2007. Cells deficient in the FANC/BRCA pathway are hypersensitive to plasma levels of formaldehyde. *Cancer Research,* 67(23):11117-22.

Roush, G.C., Walrath, J., Stayner, L.T., Kaplan, S.A., Flannery, J.T., Blair, A.1987. Nasopharyngeal cancer, sinonasal cancer, and occupations related to formaldehyde: a case-control study. *Journal of the National Cancer Institute,* 79(6):1221-4 [abstract].

Rumchev, K.B., Spickett, J.T., Bulsara, M.K., Phillips, M.R., and Stick S.M. 2002. Domestic exposure to formaldehyde significantly increases the risk of asthma in young children. *European Respiratory Journal,* 20:403–408.

Salthammer, T., Mentese, S., and Marutzky, R. 2010. Formaldehyde in the indoor environment. *Chemical Reviews,* 110(4):2536-72.

Schmid, E., Göggelmann, W., and Bauchinger, M. 1986. Formaldehyde-induced cytotoxic, genotoxic and mutagenic response in human lymphocytes and Salmonella typhimurium. *Mutagenesis,* 1(6):427-31.

Schmid, O., and Speit, G. 2007. Genotoxic effects induced by formaldehyde in human blood and implications for the interpretation of biomonitoring studies. *Mutagenesis,* 22(1): 69-74.

SCOEL (Scientific Committee on Occupational Exposure Limit Values). 2008. *Recommendation from the Scientific Committee on Occupational Exposure Limits for Formaldehyde.* SCOEL/SUM/125, European Commission.

http://ec.europa.eu/social/keyDocuments.jsp?policyArea=&type=0&country=0&year=0&advSearchKey=formaldehyde&mode=advancedSubmit&langId=en (consulted on 17th of November, 2013).

Sellakumar, A.R., Snyder, C.A., Solomon, J.J. and Albert, R.E. 1985. Carcinogenicity of formaldehyde and hydrogen chloride in rats. *Toxicology applied Pharmacology,* 81, 401–406

Shaham, J., Bomstein, Y., Meltzer, A., and Ribak, J. 1997. DNA-protein crosslinks and sister chromatid exchanges as biomarkers of exposure to formaldehyde. *International Journal of Occupational and Environmental Health,* 3: 95-104.

Shaham, J., Bomstein, Y., Gurvich, R., Rashkovsky, M., and Kaufman, Z. 2003. DNA–protein crosslinks and p53 protein expression in relation to occupational exposure to formaldehyde. *Occupational and Environmental Medicine,* 60:403–409

Soffritti, M., Maltoni, C., Maffei, F., and Biagi, R. 1989. Formaldehyde: an experimental multipotential carcinogen. *Toxicology Industrial Health,* 5(5):699-730.

Soffritti, M., Belpoggi, F., Lambertin, L., Lauriola, M., Padovani, M., and Maltoni, C.2002. Results of long-term experimental studies on the carcinogenicity of formaldehyde and acetaldehyde in rats. *Annals of New York Academy of Sciences,* 982:87-105.

Speit, G. 2006. The implausibility of systemic genotoxic effects measured by the comet assay in rats exposed to formaldehyde. *Journal of Proteome Research,* 5(10):2523-4.

Speit, G., Schmid, O., Neuss, S., and Schütz, P. 2008. Genotoxic effects of formaldehyde in the human lung cell line A549 and in primary human nasal epithelial cells. *Environmental Molecular Mutagenesis,* 49(4):300-7.

Suruda, A., Schulte, P., Boeniger, M., Hayes, R.B., Livingston, G.K. and Steenland, K. 1993. Cytogenetic effects of formaldehyde exposure in students of mortuary science. *Cancer Epidemiology, Biomarkers and Prevention,* 2: 453-460.

Takahashi, S., Tsuji, K., Fujii, K., Okazaki, F., Takigawa, T., Ohtsuka, A., Iwatsuki, K. 2007. Prospective study of clinical symptoms and skin test reactions in medical students exposed to formaldehyde gas. *Journal of Dermatology,* 34(5):283–289.

Tang, X., Bai, Y., Duong, A., Smith, M.T., Li, L., Zhang, L. 2009. Formaldehyde in China: Production, consumption, exposure levels, and health effects. *Environmental International,* 35(8):1210-24.

Teng, S., Beard, K., Pourahmad, J., Moridani, M., Easson, E., Poon, R., and O'Brian PJ. 2001. The formaldehyde metabolic detoxification enzyme systems and molecular cytotoxic mechanism in isolated rat hepatocytes. *Chemico-Biological Interactions,* 130-132: 285-296.

Thompson, C.M., and Grafström, R.C. 2009. Commentary: mechanistic considerations for associations between formaldehyde exposure and nasopharyngeal carcinoma. *Environmental Health,* 8:53.

Vaughan, T.L., Strader, C., Davis, S., and Daling, J.R. 1986. Formaldehyde and cancers of the pharynx, sinus and nasal cavity: I. Occupational exposures. *International journal of cancer,* 38(5):677-83 [abstract].

Vaughan, T.L., Stewart, P.A., Teschke, K., Lynch, C.F., Swanson, G.M., Lyon, J.L., and Berwick, M. 2000. Occupational exposure to formaldehyde and wood dust and nasopharyngeal carcinoma. *Occupational and environmental medicine, 57*(6):376-84.

Viegas, S., Ladeira, C., Nunes, C., Malta-Vacas, J., Gomes, M., Brito, M., Mendonca, P., and Prista, J. 2010. Genotoxic effects in occupational exposure to formaldehyde: A study in anatomy and pathology laboratories and formaldehyde-resins production. *Journal of Occupational Medicine and Toxicology,* 5(25):1-8.

Walrath J, Fraumeni JF, Jr. 1983. Mortality patterns among embalmers. *International Journal Cancer,* 31(4): 407-411[abstract].

Wang, B., and Liu, D.D. 2006. *Detection of formaldehyde induced developmental toxicity assessed with single cell gel electrophoresis.* Fen Zi Xi Bao Sheng Wu Xue Bao 39(5):462-6 [abstract].

Wang, W., Xu, J., Xu, L., Yue, B., and Zou, F. 2007. The instability of (GpT)n and (ApC)n microsatellites induced by formaldehyde in Escherichia coli. *Mutagenesis,* 22(5): 353-357.

Wang, M., Cheng, G., Balbo, S., Carmella, S.G., Villalta, P.W., Hecht, S.S. 2009. Clear differences in levels of a formaldehyde-DNA adduct in leukocytes of smokers and nonsmokers. *Cancer Research,* 69(18):7170-4.

Wang, Y., Yang, H., Li, L., Wang, H., Xia, X., and Zhang, C. 2012. Biomarkers of chromosomal damage in peripheral blood lymphocytes induced by polycyclic aromatic hydrocarbons: a meta-analysis. *International archives of occupational and environmental health* 85(1):13-25.

West S, Hildesheim A, Dosemeci M. 1993. Non-viral risk factors for nasopharyngeal carcinoma in the Philippines: results from a case-control study. *International journal of cancer,* 55(5): 722-727 [abstract].

WHO (World Health Organization). 2000. Air Quality Guidelines for Europe, 2nd edition. *WHO Regional Publications,* European series No. 91, Copenhagen, 273 pp.

Ye, X., Yan, W., Xie, H., Zhao, M., Ying, C. 2005. Cytogenetic analysis of nasal mucosa cells and lymphocytes from high-level long-term formaldehyde exposed workers and low-level short-term exposed waiters. *Mutation Research,* 588: 22-27.

Ye, Z., and Song, H. 2005. Glutathione s-transferase polymorphisms (GSTM1, GSTP1 and GSTT1) and the risk of acute leukaemia: a systematic review and meta-analysis. *European Journal Cancer,* 41(7):980-9.

Yu L.Q., Jiang S.F., Leng S.G., He F.S., and Zheng Y.X. 2005. Early genetic effects on workers occupationally exposed to formaldehyde. *Chinese Journal of Preventive Medicine,* 39(6): 392-5 [abstract].

Zhang, L., Steinmaus, C., Eastmond, D.A., Xin, X.K., and Smith, M.T. 2009. Formaldehyde exposure and leukemia: a new meta-analysis and potential mechanisms. *Mutation Research,* 681(2-3):150-68.

Zhang, L., Freeman, L.E., Nakamura, J., Hecht, S.S., Vandenberg, J.J., Smith, M.T., and Sonawane, B.R. 2010a. Formaldehyde and leukemia: epidemiology, potential mechanisms, and implications for risk assessment. *Environmental and Molecular Mutagenesis,* 51(3):181-91.

Zhang, L., Tang, X., Rothman, N., Vermeulen, R., Ji, Z., Shen, M., Qiu, C., Guo, W., Liu, S., Reiss, B., Freeman, L.B., Ge, Y., Hubbard, A.E., Hua, M., Blair, A., Galvan, N., Ruan, X., Alter, B.P., Xin, K.X., Li, S., Moore, L.E., Kim, S., Xie, Y., Hayes, R.B., Azuma, M., Hauptmann, M., Xiong, J., Stewart, P., Li, L., Rappaport, S.M., Huang, H., Fraumeni, J.F. Jr, Smith, M.T., and Lan, Q. 2010b. Occupational exposure to formaldehyde, hematotoxicity, and leukemia-specific chromosome changes in cultured myeloid progenitor cells. *Cancer Epidemiology, Biomarkers and Prevention*, 19(1):80-8.

Zhong, W., and Hee, S.Q. 2004. Quantitation of normal and formaldehyde-modified deoxynucleosides by high-performance liquid chromatography/UV detection. *Biomedical Chromatography,* 18(7):462-9.

In: Formaldehyde
Editor: Amy Patton
ISBN: 978-1-63482-412-5

Chapter 2

FORMALDEHYDE IN THE SYNTHESIS OF RESORCINOL-FORMALDEHYDE CARBON GELS

Natalia Rey-Raap, Ana Arenillas and J. Angel Menéndez
Instituto Nacional del Carbón, CSIC, Oviedo, Spain

ABSTRACT

Carbon gels are porous materials obtained from the polymerization of resorcinol with formaldehyde in the presence of a solvent. In this reaction, formaldehyde is the reagent responsible for the crosslinking that takes place during the polymerization and, hence, it plays an important role in determining the final porous structure of these materials.

This chapter goes deeply into the way in which formaldehyde modifies the final porous properties of carbon gels and how its influence is affected by other variables involved in the synthesis process. In general terms, an increase in the concentration of formaldehyde shifts the chemical equilibrium towards the products, thereby increasing the reaction rate. This implies that clusters are formed and grow faster. On the other hand, an excess of formaldehyde also leads to more interconnections between clusters, resulting in a materials with highly branched structures. However, the effect of formaldehyde may be modified by the effect of other reagents. For example, an increase in the amount of solvent hinders the crosslinkages produced during the reaction. Similarly, the initial pH of the precursor solution also influences the effect of formaldehyde, since acidic conditions favour the addition of formaldehyde among the resorcinol anions. The use of surfactants also affects the normal course of the sol-gel reaction. Commonly, these additives are either ionic or non-ionic and, depending on their chemical load, their effect on the formation of the porous structure will be different. Generally, anionic surfactants create repulsive forces between resorcinol anions while cationic surfactants generate attractive forces. Depending on the method of preparation of the precursor solution, non-ionic surfactants are either incorporated into the structure of the gel or remain as micelles, leaving voids when the surfactant is removed.

Generally, the formaldehyde used in the synthesis of carbon gels (37 wt. % in water) is stabilized by methanol. The role of the methanol is to prevent the polymerization of formaldehyde by forming hemiacetal groups. The concentration of methanol in the solution depends largely on the manufacturer and is usually set between 0.6 and 15 wt.%.

This variation in the concentration of methanol also affects the influence of formaldehyde on the polymerization reaction of carbon gels since the formaldehyde needs to be in its free form in order to react with resorcinol. Solutions that contain low concentrations of methanol contain larger amounts of free formaldehyde molecules and therefore, materials with greater porosity are obtained.

CARBON GELS: SYNTHESIS AND APPLICATIONS

Inspired by the synthesis of inorganic silica gels, whose method of preparation and properties have made them highly versatile materials suitable for multiple industrial applications [1], Pekala and co-workers produced the first organic gel by applying the sol-gel methodology in 1989 [2]. This methodology consists in provoking a chemical reaction in which a liquid condenses gradually from a stable suspension of colloidal solid particles (sol) to create a tridimensional porous network formed by polymeric chains in a liquid medium (gel). In the case of organic gels, from which carbon gels are obtained, the sol-gel reaction occurs between hydroxylated benzenes and aldehydes in the presence of a solvent [2-7]. The gel resulting from this reaction goes through a curing stage, in which the material becomes harder due to the simultaneous occurrence of crosslinkages between polymeric chains and the partial evaporation of the solvent [1, 8, 9]. Because the removal of the solvent is not completed during curing, a drying stage is needed in order to completely eliminate the solvent embedded inside the polymeric structure [1, 10]. The dried gel is then stabilized by means of thermal treatments in which oxygenated and hydrogenated groups are removed, resulting in a carbonaceous material known as carbon gel [9, 11].

Figure 1 shows the main stages involved in the synthesis of carbon gels by means of the sol-gel method: the sol-gel reaction and curing, drying and carbonization. Each of these steps plays an important role in determining the final properties of the resultant materials and therefore, each one is reviewed below in more detail.

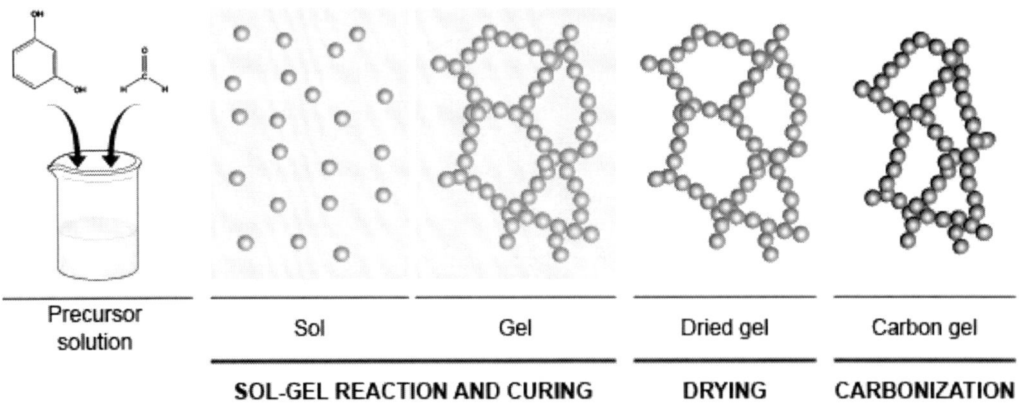

Figure 1. Main stages involved in the synthesis process of carbon gels.

Sol-Gel Reaction and Curing

The synthesis of carbon gels begins with the preparation of an aqueous solution containing hydroxylated benzene, aldehyde and solvent and, when necessary, a fourth compound, known as catalyst, which regulates the pH and catalyzes the reaction [9]. The nature and concentration of these reagents, the pH of the precursor solution, the incorporation of additives and the heating method all play a key role in the synthesis of organic gels, i.e., in the sol-gel reaction and curing stages [12].

The choice of a suitable reagent for synthesizing organic gels is essential for the sol-gel reaction to take place. Several precursors can be employed to develop carbon gels, as is shown in Figure 2. Phenol [4, 5, 13-15], cresol [6, 16-18], cellulose [19] or tannins [20-22] have been used as the hydroxylated benzene, though resorcinol is probably the most commonly used for the synthesis of carbon gel [2, 8, 9, 23, 24]. For aldehydes, formaldehyde is the most widely employed reagent because of its importance for the development of the polymer structure [25]. However, some studies have also resorted to furfural [3, 26-28] or hexamethylenetetramine (HMTA) [5, 29].

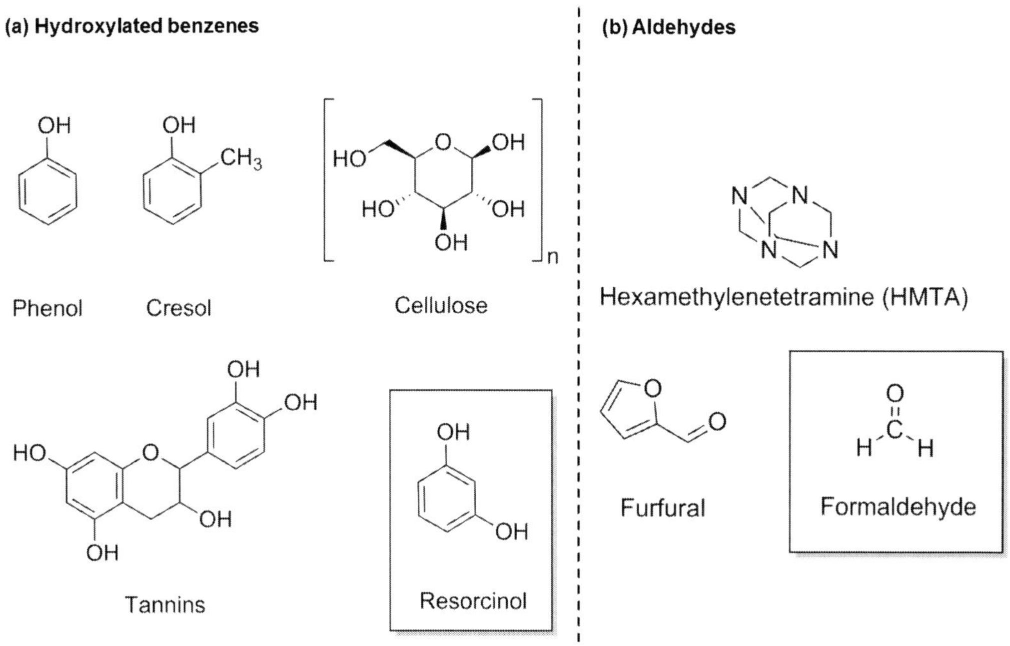

Figure 2. The most common reagents used for the synthesis of carbon gels as hydroxylated benzene (a) and as aldehyde (b).

The reaction medium is also an important factor in the synthesis of carbon gels. Several solvents are available including deionized water [23, 24, 30, 31], acetone [32], methanol [33-35] or ethanol [36]. Each of these compounds has a different surface tension and therefore, their behavior during the curing and drying stages will result in different pore structures. Although water exhibits the highest surface tension value giving rise to a further collapse of the polymeric structure when the solvent evaporates, it is the least expensive reaction medium and therefore, the most preferred for synthesizing carbon gels. The nature of the catalyst may also influence the sol-gel reaction because of their different capacities of polarization [37].

Catalysts can be acidic such as perchloric acid [38, 39], nitric acid [40], citric acid [41] and acetic acid [42] or basic such as alkali carbonates [7, 30, 43-47] or sodium hydroxide [13, 31, 36, 48-51]. However, basic catalysts are the most frequently used [9, 52].

Of all the possible reagents available, the most commonly employed for synthesizing carbon gels are resorcinol and formaldehyde [2, 12, 23, 24, 31, 53, 54]. Hence, from now on, all references regarding the precursors of carbon gels will assume the involvement of these two reagents.

It is important to emphasize that the type of reagent is not the only important factor in the synthesis of carbon gels. The concentration of the reagents also has a strong influence on the final properties of these materials [9, 10, 23, 52, 55]. These concentrations are related to the molar ratio between the resorcinol and the formaldehyde [23, 25, 55, 56], the pH of the precursor solution [2, 7, 10, 24, 37, 44, 52, 57] and the dilution ratio [10, 23, 37, 44, 48, 57-59].

The effect of the molar ratio between the resorcinol and the formaldehyde is determined by the amount of formaldehyde. This reagent is responsible for the crosslinking between clusters that takes place during the sol-gel reaction and curing stage [25]. By a way of introduction it may be said that a high concentration of formaldehyde could result in materials with greater porosity by choosing suitable synthesis conditions [55]. The influence of the pH of the precursor solution on the final properties of the carbon gels has been one of the most widely studied variables and its value depends on the amount of catalyst added [2, 7, 23, 37, 44, 52, 57, 60]. An increase in the pH value favours the addition of formaldehyde, leading to the formation of highly branched aggregates and, in turn, to smaller and more interconnected clusters [7, 23, 37, 44, 52, 57, 60]. The dilution ratio, commonly defined as the molar ratio between the total solvent and reagent, can modify the behavior of the formaldehyde in the sol-gel reaction. The effects of varying the degree of dilution are diverse and depend largely on the conditions chosen for the preparation of the precursor solutions and on the methods employed for the synthesis and drying stages. Generally speaking an increase in the amount of water will results in a structure with a few weakly branched clusters [48]. The mechanical strength of these materials is low, and therefore, during the drying stage they will undergo further contraction, leading to materials with a smaller pore volume [7, 23, 48, 57, 58].

Although modifying the concentration of the reagent is the most common way to change the pore structure of carbon gels, the addition of additives can also affect the normal course of the sol-gel reaction, by altering the formation of the polymeric structure [61]. The additives most commonly used in the synthesis of carbon gels are surfactants [61, 62]. In general, it may be said that there are three fundamental types of surfactant: anionic, cationic and non-ionic [61-65]. The nature and concentration of each surfactant alters the behavior of formaldehyde in the sol-gel reaction and, in turn, the final shape of the porous structure [61, 62].

Once the reagents involved in the sol-gel process have been selected and mixed in the appropriate proportions, the precursor solution of the carbon gel is subjected to a heating process in order to initiate the sol-gel reaction, which will yield a stable organic wet gel. The process most frequently used to carry out the sol-gel reaction and curing involves heating the precursor solutions in a conventional electric furnace at a temperature not exceeding 100 ºC [2, 6, 7, 31, 66]. The synthesis of organic wet gel by means of conventional heating requires several days to produce the final material, resulting in a less cost-competitive material than activated carbon, which is the one preferred for most applications because of its low production cost [31, 53]. Hence, the implantation of the synthesis of carbon gels at industrial scale entails reducing the cost of the manufacturing process, in order to produce more competitive materials than activated carbons. Researchers in this field are directing their effort towards the development of faster and cheaper heating methods such as ultrasonic and microwave radiation [26, 31, 47].

Ultrasonic radiation has been found to be a helpful strategy for increasing reaction rates and product yields and for shortening the reaction time required [67]. However, so far this heating method has only been used for the gelation of the material, the curing and drying stages being performed in a conventional furnace [67].

Microwave radiation as a heating method for the synthesis of carbon gels was applied for the first time in 2006 [68]. Initially, the application of microwave heating for the synthesis of carbon gels was directed at some specific stage of the synthesis process, e.g., the gelation and curing stages which were then followed by drying in an electrical oven [69] or just the drying stage [53, 68]. In all cases, the organic gels obtained exhibited a well-developed porous structure and both time and costs were greatly reduced. Nowadays, microwave heating is used to carry out all three stages of the production process (the sol-gel reaction, curing and drying), resulting in porous materials with similar properties to those obtained by conventional methods [31]. Thus, it can be said that microwave radiation is as a cost-effective technology that allows the synthesis of organic gels on a large scale via a simple and rapid procedure, yielding very competitive and high value-added materials.

Drying

Whatever the heating method used for synthesizing organic gels, the material obtained must be dried so that the solvent embedded in the polymeric structure can be completely removed. To eliminate the solvent, not only is it necessary take into account the type of solvent added but also the solution in which the formaldehyde is dissolved. Generally, the formaldehyde employed to synthesize carbon gels (37 wt. % in water) is stabilized by methanol. The role of the methanol is to prevent the polymerization of the formaldehyde by forming hemiacetal groups, which are molecules of stabilized formaldehyde [35]. The concentration of methanol in the solution depends largely on the manufacturer and, depending on this concentration and on the drying method used, the surface tension at the liquid-vapor interface can be altered and consequently, different degree of shrinkage can occur. Accordingly, not only are the nature and concentration of the reagent key factors in the synthesis of carbon gels but also the method of solvent removal [59].

Three different types of drying method can be found in the literature, supercritical drying, freeze-drying and subcritical drying, each of which is carried out under different process conditions, as shown in Figure 3.

The most widely used drying method for the synthesis of organic gels is supercritical drying [13, 28, 37, 47, 70]. In this method the solvent is replaced by CO_2 which is subsequently disposed of in a process carried out at high pressure [28, 37, 70]. The main problem with this method is that if water is used as solvent it must be replace by an organic solvent prior to its replacement by CO_2, due to the solubility of CO_2 in water [71]. Alternatively, the CO_2 replacement step can be eliminated by directly removing the organic solvent at high pressure and high temperature [72]. However, in this case the structure undergoes further collapse, leading to materials with less porosity [72]. The gels obtained by applying this drying method are known as aerogels and their properties make them very promising materials for a wide range of applications. Notwithstanding, it is a laborious, time-consuming and expensive method and, therefore, difficult to carry out at an industrial scale [24].

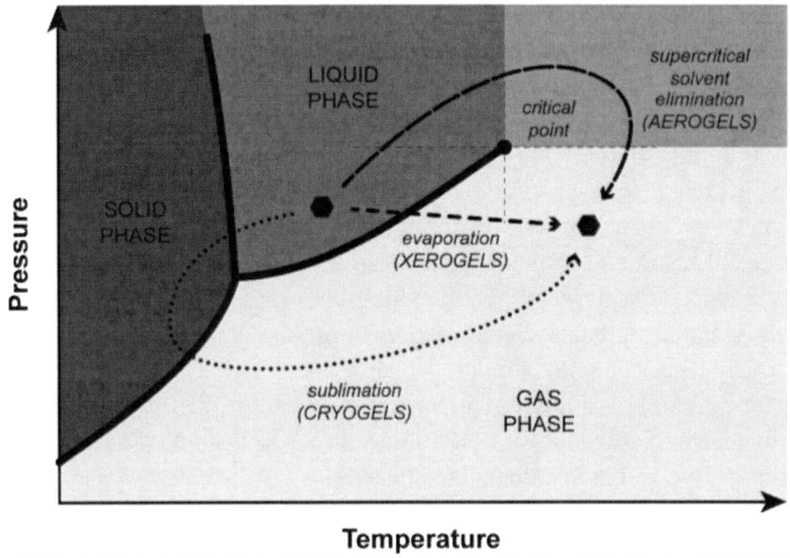

Figure 3. Scheme of the different drying conditions used in the synthesis of organic gels.

Cryogenic drying involves freezing the solvent that is embedded in the polymeric structure and its subsequent removal by sublimation [59]. If water is used as solvent, it must be replaced by an organic solvent in order to prevent the formation of megalopores [30, 59] and changes in the density of the gel after freezing [59]. The materials obtained by this drying technique are known as cryogels and generally exhibit a high mesoporosity [59]. Although it is somewhat cheaper than the supercritical drying method, the need to use organic solvents or to have the water replaced by an organic solvent entails an increase in time and costs, making the process unfeasible at industrial scale.

The third drying method, which results in materials called xerogels, is based on the direct evaporation of the solvent [59, 70] and hence, the drying temperature will greatly depends on the type of reagent used as solvent. However, unlike the other two drying methods, the subcritical conditions lead to the formation of a liquid-vapor interface and, when the solvent evaporates high surface tensions emerge that may cause the collapse of the polymeric structure [10, 24, 54, 59, 70].

Nevertheless, although shrinkage occurs during drying, several studies have reported that carbon xerogels with a well-developed porous structure can be obtained if appropriate process conditions are chosen [7, 24, 31, 53]. Furthermore, it is the cheapest, easiest and fastest method of drying and therefore, economically viable at industrial scale [36]. Hence, carbon xerogels with similar characteristics to those of aerogels and cryogels can be obtained but by via of a more competitive method.

Carbonization

Once dried organic gels have been obtained, carbonization must be applied in order to produce the carbon gels. The main aims of this process are first to stabilize the material and, secondly to increase the micropore volume and surface area and also to widen the narrow pores created during the process of synthesis [7, 8, 36, 73].

Carbonization, the most commonly used thermal treatment, involves heating the sample in a non-reactive atmosphere [1]. Time, temperature and the nature and flow rate of the carrier gas are the parameters that have the greatest effect on the final properties of the carbon gel [11, 73-75]. Temperatures between 600 and 1000 ºC are generally used with slow heating ramps to prevent excessive shrinkage of the structure, as a result of the removal of the gases produced inside the material [73, 75]. Nitrogen, helium or argon are used as carrier gas and constant and moderate flows are selected according to the amount of sample and the size of the tube furnace.

Several authors have reported that the carbonization process involves the breaking up of C-O bonds at low temperatures and the breaking up of C-H bonds at high ones [76, 77]. The breakage of bonds leads to the rearrangement of molecules and more stable structure [58]. Furthermore, according to some published works [78], the walls separating individual micropores break, resulting in the enlargement of narrow micropores [78]. On the other hand, the condensation produced during the carbonization process causes a slight shrinkage of the polymeric structure and, consequently, the size of the pores decreases [60]. Both phenomena contribute to an increase in the microporosity during carbonization.

It is well known that some applications require materials with a high amount of microporosity, which cannot be achieved by carbonization alone. Consequently, another thermal treatment known as activation need to be applied to organic gels in order to increase the microporosity of these materials [46, 78]. The activation process can be carried out either during or after carbonization, resulting in materials with different properties. The temperature and time of activation, the activating agents and precursors are some of the variables that need to be optimized in order to obtain well-developed activated carbons [75, 78]. Basically, there are two types of activation methods: physical activation and chemical activation [78]. Physical activation is the most widely used method. With this thermal treatment, the material is heated under a reactive gas flow. Gases such as air, steam or CO_2 or a combination of them are generally used [75]. Chemical activation involves mixing the sample with an activating agent and then increasing the temperature. This method requires a subsequent washing step to remove the unreacted chemical agent and any possible residual products from the reaction [46]. Examples of the activating agents used are phosphoric acid, zinc chloride, alkali metal carbonates and alkali hydroxides [12, 46, 79]

Applications

Carbon gels are very versatile materials whose polymeric structure can be designed to fit the requirements of a specific application by selecting the appropriate process conditions. Bearing in mind the shape of the polymeric structure of carbon gels, the porosity of these materials can be defined as: i) voids within polymeric clusters with a width of less than 2 nm known as micropores and [10], ii) voids between polymeric clusters, which are called mesopores when their width is in the 2-50 nm range and macropores when the size of voids are wider than 50 nm [10]. A schematic representation of the porous structure of RF xerogels is shown in Figure 4.

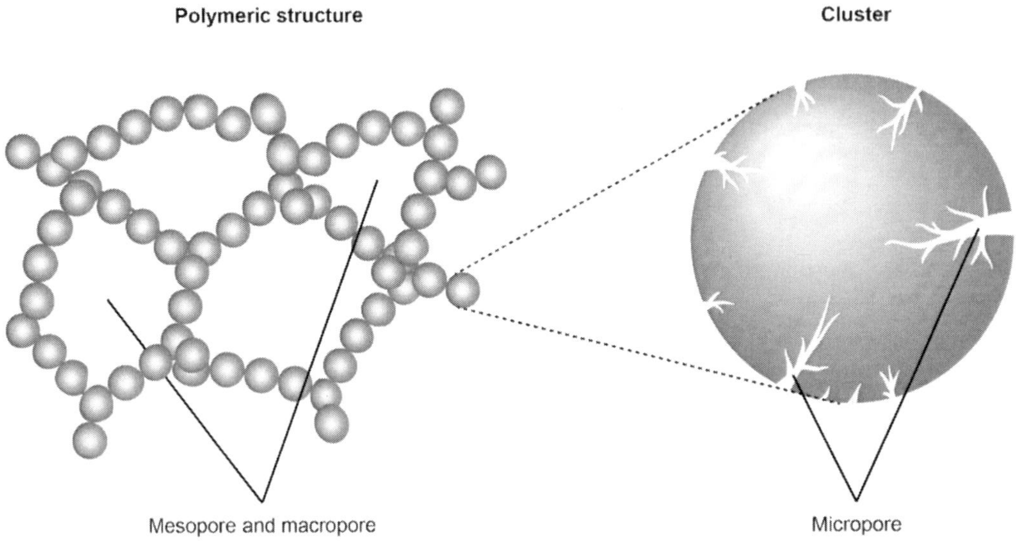

Figure 4. Schematic representation of the porous structure of carbon gels.

Depending on the shape of the porous structure formed, carbon gels will differ in surface area, volume and pore size, density, electrical and thermal conductivity, chemical composition, etc. The right combination of these properties makes carbon gels perfect candidates for various applications such as adsorption [11, 23, 52], hydrogen storage [28, 80], electrode material for supercapacitors [51, 81-84] or lithium batteries [85, 86], catalyst support [10, 23], etc.

In dynamic adsorption processes, diffusion limitations can be minimized by choosing a carbon gel with pores large enough to overcome mass transfer limitations [87]. A high hydrogen storage capacity can be achieved by using carbon gels with a highly developed microporosity and an appropriate micropore size distribution [80]. When carbon gels are to be used as electrode material, in addition to microporosity, mesopores are required in order to facilitate a good electrolyte diffusion [69, 81, 82]. In the case of catalyst supports carbon gels need to have a porous structure that allows an adequate support and dispersion of the metal particles added in the catalytic reactions [10]. Accordingly, the porous and chemical properties of the carbonaceous material should be designed with a view to the type of metal that is going to be used and the type of reaction in which it is going to be employed.

Role of Formaldehyde in the Synthesis of Carbon Xerogels

Controlling the porous structure of resorcinol-formaldehyde (RF) xerogels is essential for tailoring the properties of these materials to fit the requirements of a specific application. As mentioned in the previous section, there are many variables that need to be controlled for tailoring the final properties. These variables are related to the conditions selected in each stage of the production process: sol-gel reaction and curing, drying and carbonization. Although all of these stages are key factors for designing the final porous properties of carbon gels, this section will focus principally on the synthesis of organic gels (sol-gel reaction, curing and drying), since this is where formaldehyde, the main protagonist of this chapter, plays an important role.

Formaldehyde is the most widely used aldehyde for performing the sol-gel reaction in order to obtain RF xerogels [55]. Most of the published studies on the synthesis of RF gels have employed the stoichiometric molar ratio between the resorcinol and the formaldehyde (i.e., an R/F molar ratio value of 0.5) [7, 8, 11, 24, 39, 46, 48, 53, 59, 78]. However, some studies report that it is possible to modify the concentration of formaldehyde altering the sol-gel reaction and as a consequence the porous structure of the gels [23, 25, 55, 58].

The role of the formaldehyde in the formation of the porous structure can be explained by the polymerization reaction between resorcinol and formaldehyde. According to several published works [2, 7, 52] the polymerization reaction involves two main stages: i) an addition reaction in which hydroxymethyl derivatives of resorcinol are formed and, ii) a condensation reaction in which the hydroxymethyl derivatives are linked by methylene and ether bonds.

All the steps involved in the polymerization reaction between the resorcinol and formaldehyde are illustrated in Figure 5. First, resorcinol anions are formed due to the abstraction of hydroxyl hydrogen, as shown is Figure 5a [2, 88]. Resorcinol is a trifunctional benzyl compound with two hydroxyl groups in positions 1 and 3 which allows the addition of formaldehyde in positions 2, 4 and 6 [10]. This feature along with the fact that resorcinol anions, formed at the beginning of the addition reaction, are more reactive than resorcinol itself, leads to the addition of formaldehyde [10]. Primarily, the addition of formaldehyde occurs via the formation of hydroxymethyl derivatives (Figure 5b), which are the monomers necessary for polymerization to occur [10, 44].

During the addition reaction condensation also takes place. The hydroxymethyl derivatives lose OH^- groups to form benzyl-type cations, as shown in Figure 5c [61]. Each cation reacts with the benzene ring of another molecule linking themselves to each other by forming methylene and ether bonds as illustrated in Figure 5d and Figure 5e, respectively [2, 10, 44, 60, 66, 89]. As the reaction proceeds the number of bonds between the rings increases, to form the polymer backbone, which leads to cross-linked polymer clusters. These clusters aggregate and crosslink with each other to complete the sol-gel reaction and curing steps [9, 10, 25], resulting in a three-dimensional interconnected porous structure (Figure 5f). Therefore, from this reaction mechanism, it can be seen that formaldehyde is the factor responsible for the formation of clusters and crosslinkages during the polymerization reaction.

According to Feng et al. [58], an increase in the concentration of formaldehyde, i.e., a decrease in the R/F molar ratio, shifts the chemical equilibrium toward the products. This suggests that the clusters will form and grow faster, allowing the reaction medium to last longer during the nucleation stage. The fact that clusters have more time to grow gives rise to the formation of a small number of large clusters. However, an excess of formaldehyde not only accelerates the formation and growth of clusters but also creates more interconnections between them, leading to a highly branched structure of high mechanical strength [2, 58]. A small number of large clusters gives rise to a low vapor-liquid interface causing small surface tensions to appear during drying. This phenomenon together with the high mechanical strength just mentioned, prevents the collapse of the structure, resulting in materials with a low degree of shrinkage and a low density, as shown in Figure 6 [58].

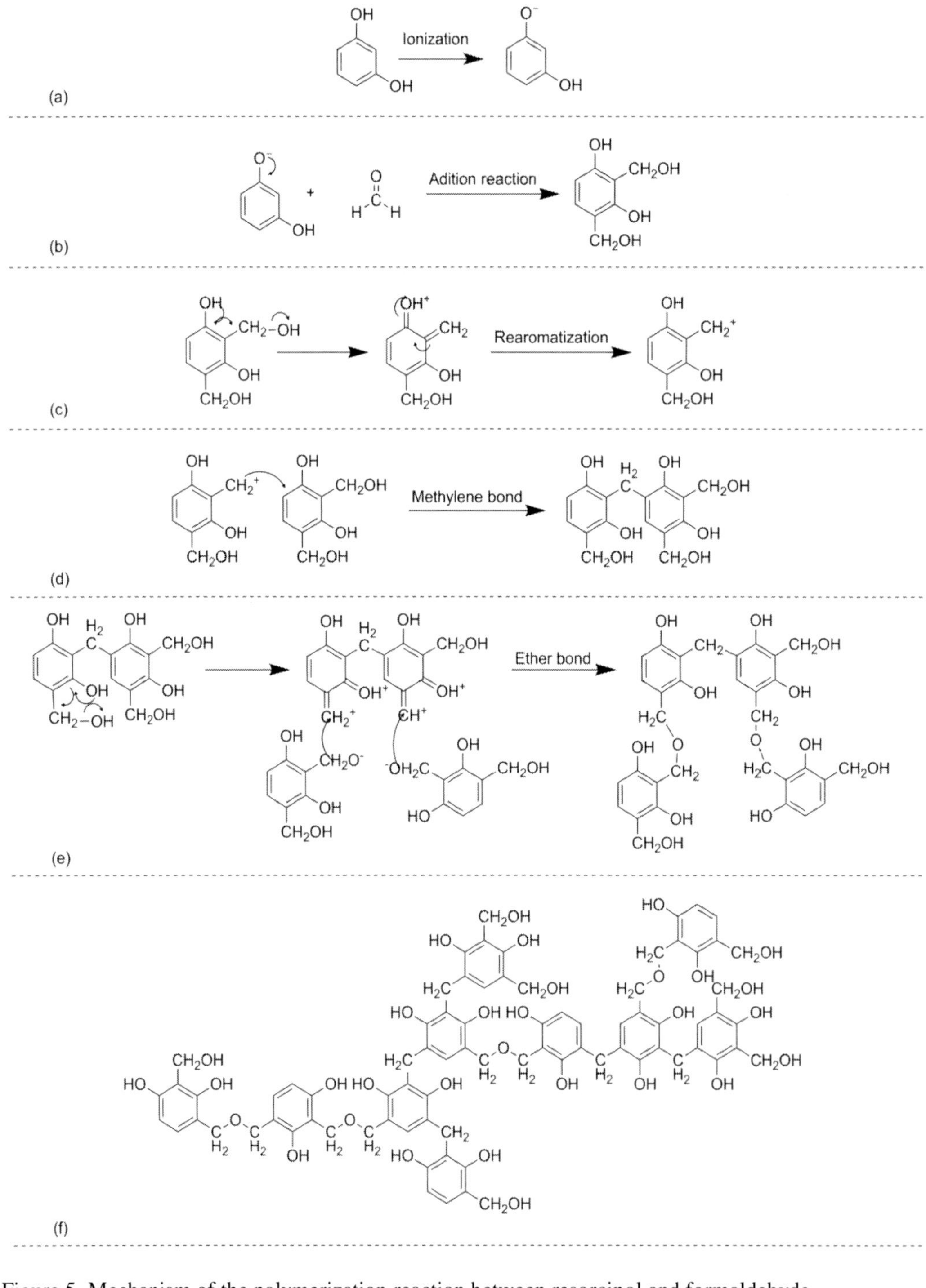

Figure 5. Mechanism of the polymerization reaction between resorcinol and formaldehyde.

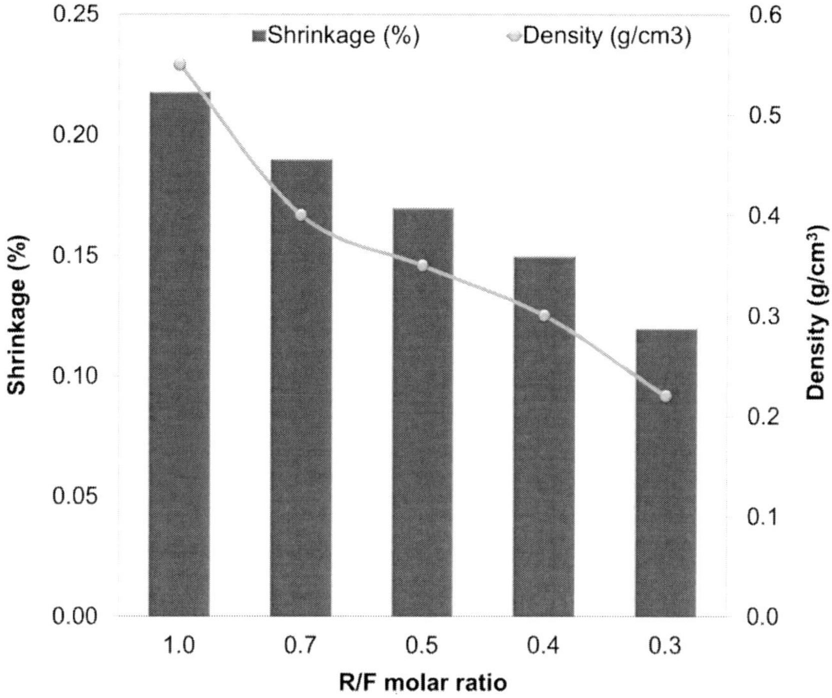

Figure 6. Percentage of shrinkage and density of carbon aerogels as a function of the R/F molar ratio. Adapted from reference [58].

On the other hand, an insufficient amount of formaldehyde (i.e., when the R/F molar ratio value is above the stoichiometric value) leads to the formation of small clusters and a weakly branched structure [25]. Small clusters give rise to a high vapor-liquid interface and as a consequence, considerable surface tensions during drying, especially when direct evaporative drying is applied. As the slightly branched structure is not able to counteract the surface tension the material shrinks, leading to the maximum compaction of the porous structure, which ends when the clusters come into contact. The compaction of the polymeric structure when the concentration of formaldehyde decreases results in a highly dense material (Figure 6).

In the literature dealing with the variation of the amount of formaldehyde to be added for the synthesis of RF gels, maximum and minimum R/F molar ratio values are proposed for obtaining well-developed porous structures. Some published works quote a value of 0.7 as the maximum limit [23, 25]. The use of concentrations of formaldehyde for below the stoichiometric value (i.e., R/F molar ratio values above 0.7) results in some cases in non-porous materials [23, 25], whilst in other cases such low values could prevent the precursor solution from reaching the gelation point [55]. However, there is considerable controversy as to the minimum R/F molar ratio value that can be used for the synthesis of RF gels. Some authors have stated that this value needs to be above 0.4 [23, 25], while other authors have reported the synthesis of well-developed porous RF gels using R/F molar ratio value of below 0.4 [39, 55, 56]. These differences are due to the fact that the formation, growth and crosslinking of the clusters not only depend on the concentration of formaldehyde and resorcinol but also on the concentration of the other reagents used to prepare the precursor solution: i.e., water and the catalyst [55]. The concentrations of these compounds are reported in the literature to be related to the dilution ratio [10, 23, 37, 44, 48, 57-59] and the pH of the precursor solution [2, 7, 10, 24, 37, 44, 52, 57], respectively.

pH of the Precursor Solutions

The pH of the precursor solution is given by the concentration of the catalyst [59]. In fact, in many of the studies that can be found in the literature the pH is defined by the R/C parameter which is the molar ratio between the resorcinol and the catalyst [8]. For any given concentration of resorcinol, the higher the concentration of the catalyst, the lower the value of the R/C ratio.

Initially, the precursor solution comprising resorcinol, formaldehyde and water has a pH value of about 3-4 [48]. Base catalyst can be added until maximum pH value is obtained. However, many studies that have dealt with the dependence of porosity on the initial pH have reported that precursor solutions with a pH value above 7 are hardly able to produce porous material [7, 48]. Consequently, most studies only consider a pH variation value of 3-7 as viable [48].

The effect of the pH of the precursor solution on the final properties of the RF xerogel has been one of the most studied variables [7, 9, 41]. Several works published in the literature have reported that both the volume and the size of the voids between the clusters are modified by the pH of the precursor solution whereas the micropore volume is unaffected [7]. Figure 7 shows the volumes of micropores, mesopores and macropores of carbon xerogels. As can be seen, the microporosity is independent of the R/C ratio and depends exclusively on the carbonization process [46]. It can also be appreciated from Figure 7 that the size and volume of the voids between the clusters (mesoporosity and macroporosity) depends largely on the R/C ratio of the precursor solution [46]. Thus, by means of small variations in the pH value, RF xerogels can be obtained corresponding to micro-macroporous materials, micro-mesoporous materials and even microporous materials.

The effect of the pH on the porosity of RF xerogels can be explained taking into account the polymerization reaction between resorcinol and formaldehyde shown in Figure 5. It can be seen from Figure 5a that the reaction starts with the abstraction of the hydroxyl hydrogen from the resorcinol. The addition of catalyst promotes the abstraction of hydroxyl hydrogens and in turn, the formation of resorcinol anions [7]. Therefore, increasing the concentration of catalyst (i.e., raising the pH value or lowering the R/C ratio) causes an increase in the number of resorcinol anions to which the formaldehyde is added [7]. As the number of resorcinol anions increases, the addition reaction proceeds faster. Thus, the addition reaction is favored by increasing the pH, resulting in the formation of a large number of small clusters [7, 25, 52], as shown in Figure 8. On the other hand, the addition of low concentrations of catalyst (i.e., low pH values or a high R/C ratio) leads to the formation of a small number of resorcinol anions [41]. As the number of resorcinol anions available decreases the addition of formaldehyde takes place more slowly than when more catalyst is added [8, 41, 52]. Several works suggest that in this case, it is not the addition reaction that is favored but the condensation reaction, as a results of which the reaction medium lasts longer during the nucleation step [7, 41]. Consequently, the number of clusters is small but of a large size (Figure 8).

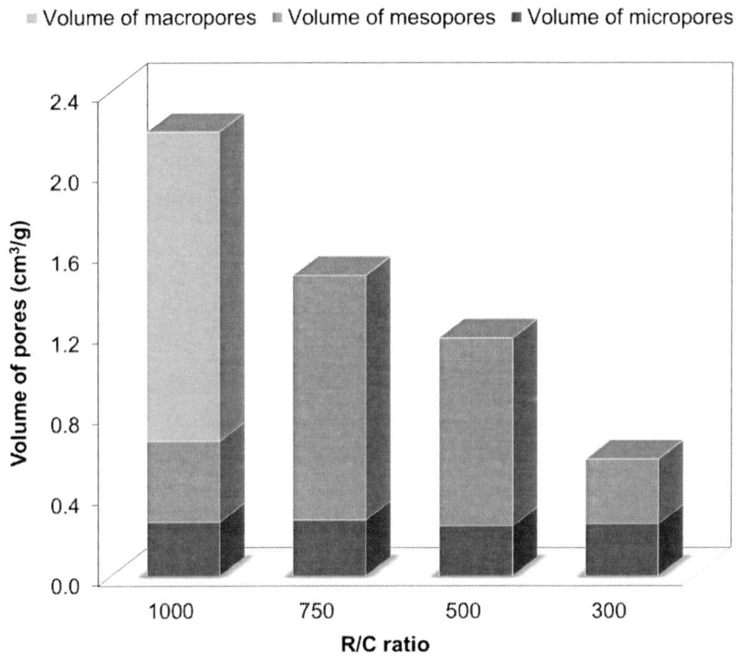

Figure 7. Volume of micropores, mesopores and macropores of carbon RF xerogels synthesized from precursor solutions with different pH values. Adapted from reference [46].

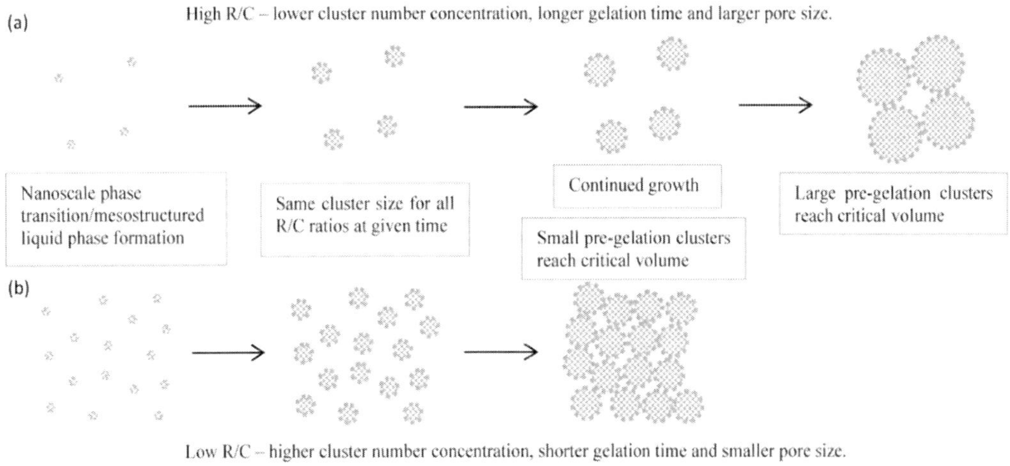

Figure 8. Schematic representation of cluster formation and growth process of resorcinol-formaldehyde gels synthesized from precursor solutions with low pH values (a) and high pH values (b). Reproduced with permission from reference [52].

Samples synthesized from precursor solutions with high pH values give rise to a greater number of clusters of small size, resulting in a more compact structure [23]. Some authors have observed that compaction of the polymeric structure occurs during the shrinkage that takes place when the material is dried [8]. During drying, porous structures formed by a large number of small clusters exhibit a high vapor-liquid interface, leading to the appearance of strong surface tensions and capillary forces [8]. Most of the formaldehyde in the precursor solution has been used in the addition reaction (i.e., in the formation of clusters) leaving less formaldehyde available for the formation of crosslinkages between the clusters. Hence, a weakly branched structure is formed. The mechanical strength of this structure is not able to counteract the surface tension produced during drying, causing the material to shrink until all the clusters are touching each other. The voids remaining between the clusters are small and therefore, materials synthesized with high pH values have a small pore size, as shown in Figure 9.

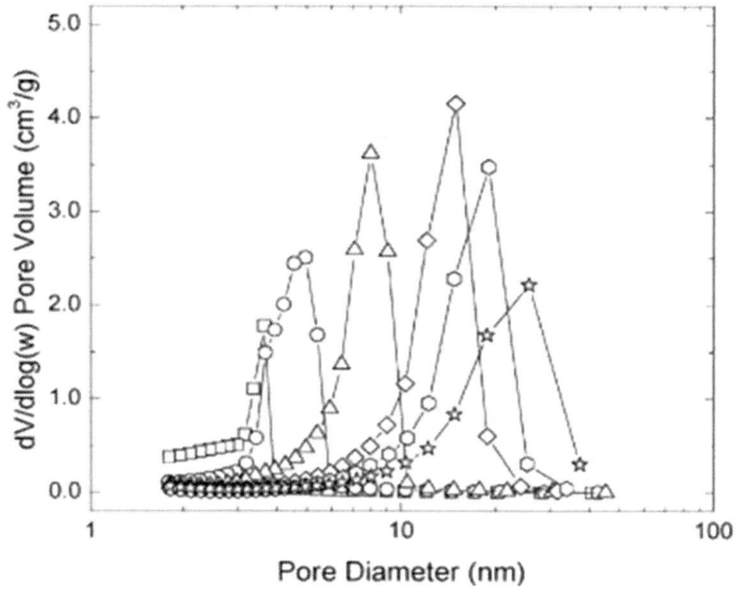

Figure 9. Pore size distribution of RF xerogels synthesized using precursor solutions with R/C ratios of (□) 100, (○) 200, (Δ) 300, (◊) 400, (⬠) 500, and (☆) 600. Reproduced with permission from reference [52].

Dilution Ratio

The dilution ratio (D) is a parameter that is directly related to the amount of solvent that is added to the precursor solution. Most studies dealing with the synthesis of RF xerogels define the dilution ratio as the molar ratio between the total solvent and reactants [7, 48]. Total solvent refers to the water and methanol contained in the formaldehyde and the deionized water that is added, whilst the term reactant refers to the resorcinol and pure formaldehyde [48]. Some studies also include the moles of catalyst in the term reactant [30, 59]. The dilution ratio has also been defined as the molar ratio between resorcinol and water [89], while other authors defined this variable as the solid content of the precursor solution, usually in terms of weight percentage [90].

Based on the most commonly used definition of dilution ratio, a remarkable number of published studies on RF gels employ a dilution ratio value of 5.7 [7, 12, 31, 53, 78, 87]. Nevertheless, it is also possible to find publications that deal with the effect of varying the total amount of solvent [23, 30, 48, 59]. The influence of the dilution ratio is somewhat controversial probably due to differences in the method of synthesis and drying used by each author. Some authors have reported that the degree of dilution does not affect the porous properties of RF aerogel synthesized by means of conventional electric furnaces [59], while other authors have found significant differences in the porous structure of RF xerogels synthesized by means of microwave heating [48]. With microwave heating the sol-gel reaction takes place much faster so that the solvent does not have enough time to evaporate and much of it remains embedded within the structure. This influences the behaviour of the formaldehyde during the polymerization reaction and drying, modifying the porous structure of the carbon xerogels. Figure 10 shows the volume of micropores, mesopores and macropores of organic RF xerogels synthesized by microwave heating. As can be seen, the micropore volume of the RF xerogels is independent of the dilution ratio, while mesoporosity and macroporosity are significantly modified when this variable is adjusted [23].

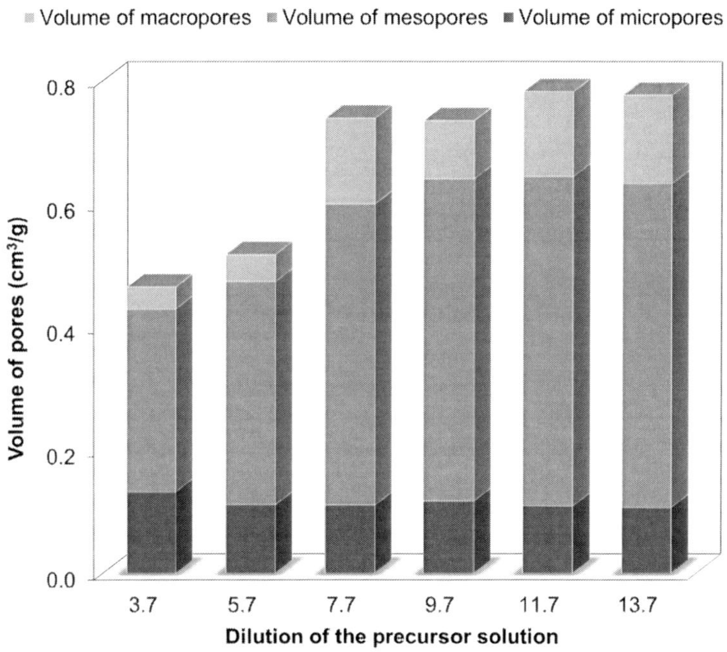

Figure 10. Volume of micropores, mesopores and macropores of organic RF xerogels synthesized by microwave heating from precursor solutions with a pH of 6.5 and different dilution ratio values.

As previously mentioned, the mesoporosity and macroporosity of RF xerogels depend on the number and size of the clusters formed during the sol-gel reaction and curing steps and on the shrinkage that occurs during drying. Therefore, the effect of the dilution ratio on the porosity of RF xerogels can be explained by the polymerization reaction between resorcinol and formaldehyde illustrated in Figure 5. According to the definition of the dilution ratio (D) given above and for the same initial volume of precursor solution, low values of D involve high concentrations of resorcinol and formaldehyde (RF). The higher the concentration of resorcinol, the greater will be the number of resorcinol anions generated, which will favour the addition of formaldehyde [55]. SEM images of samples synthesized from RF precursor solutions with different dilution ratios are shown in Figure 11. As can be seen, a decrease in the dilution ratio gives rise to the formation of a large number of small clusters. These results are in good agreement with those of a recent published work dealing with the variation of the dilution ratio [90], whose authors also reported that an increase in the speed of the addition reaction had resulted in the formation of a large number of small clusters [90].

Figure 11. SEM images of carbon xerogels synthesized from RF precursor solutions with a pH value of 5.8 and low (a) and high (b) dilution ratio values.

Increasing D values (i.e., decreasing RF content) lead to the formation of a small number of resorcinol anions. As the number of resorcinol anions decreases, the addition of formaldehyde takes place more slowly, favoring the condensation reaction and resulting in a small number of large clusters [48]. Porous structures with large clusters have a low vapor-liquid interface, which leads to small surface tensions during the drying process. Furthermore, according to some published studies [89], low concentrations of resorcinol and formaldehyde (i.e., high D) result in poorly branched clusters. This phenomenon along with the low degree of shrinkage that occurs during drying leads to less dense materials, as shown in Figure 12.

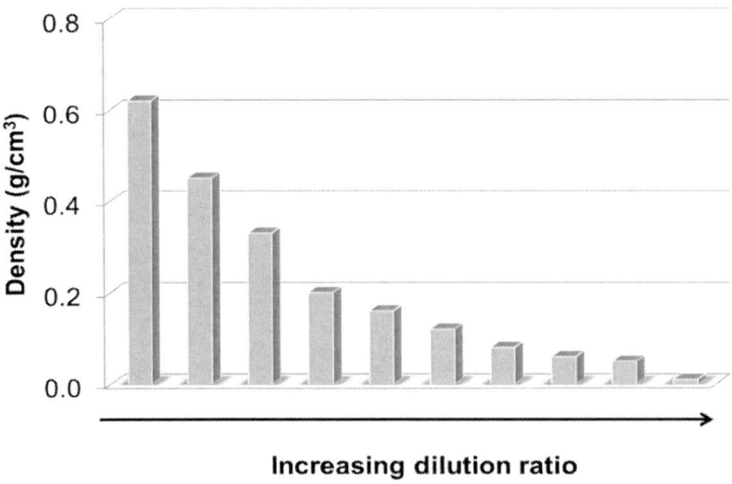

Figure 12. Evolution of the density of resorcinol-formaldehyde organic aerogel with the modification of the dilution ratio. Adapted from reference [91].

Influence of pH and Dilution Ratio on the Behavior of Formaldehyde

From the evidence presented so far it is clear that each one of the variables related to the concentrations of the reactants (R/F molar ratio, pH and dilution ratio) has some influence on the mechanism of the sol-gel reaction. Hence, these three variables are key factors in the design of the porous properties of RF gels as they can be adjusted in order to control the number and size of the clusters formed and, consequently, the number and size of the voids between them, i.e., mesoporosity and macroporosity of the materials. However, it is important to note that the effect of each of these variables may be modified by the effect of the others. In support of this idea, Rey-Raap et al. graphically showed all the combinations of pH-D from which RF xerogels could be synthesized by means of microwave heating [48]. These authors reported that when the dilution ratio was too high (i.e., the amount of RF was too low) only a few small clusters were formed [48]. This phenomenon and the fact that a large amount of water increased the distance between the clusters prevented the solution from reaching the gelation point. Accordingly, controlled synthesis of RF xerogels was only achieved in the pH range of 3.0-7.0 when the dilution ratio was fixed at below 14. Within this range, porous materials with very different properties were obtained due to the combined effect of the pH and D on the addition of formaldehyde. Figures 13a and 13b present contour plot showing the effect of the pH and dilution ratio on the volume of mesopores and macropores of RF xerogels, respectively.

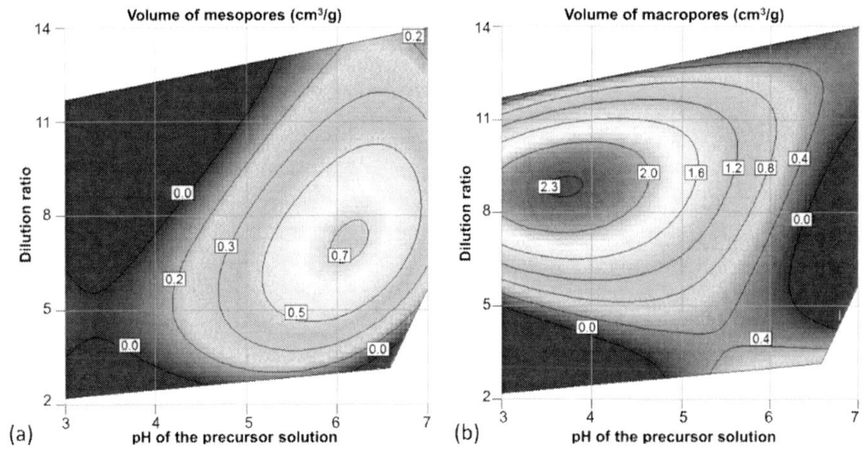

Figure 13. Contour plots showing the effect of the pH and dilution ratio on the mesoporosity (a) and macroporosity (b) of RF xerogels. Reproduced with permission from reference [48].

As can be seen in Figure 13 the samples with a pH above 5.5 are exclusively mesoporous while the sample synthesized from a precursor solution with a pH below 5.5 gives rise to macroporous materials. However, for any given pH, the highest mesoporosity and macroporosity which can be obtained for each sample is dependent on the dilution ratio used. From Figure 13, it can be concluded that within the range of mesoporosity (i.e., pH values above 5.5) the maximum pore volume is obtained by fixing the dilution ratio between 6 and 8, and in the case of macroporous samples between 8 and 10. However, as already mentioned, formaldehyde favours the condensation reaction that leads to the formation of a small number of large clusters. Some authors state that, within the range of maximum mesoporosity shown in Figure 13a, the volume of mesopores can be increased even more by lowering the R/F molar ratio size [55]. Figure 14 shows the volume of micropores and mesopores of organic RF aerogels [23]. As can be seen, the microporosity of the carbon gels is independent of the R/F molar ratio while mesoporosity depends largely on the amount of formaldehyde added.

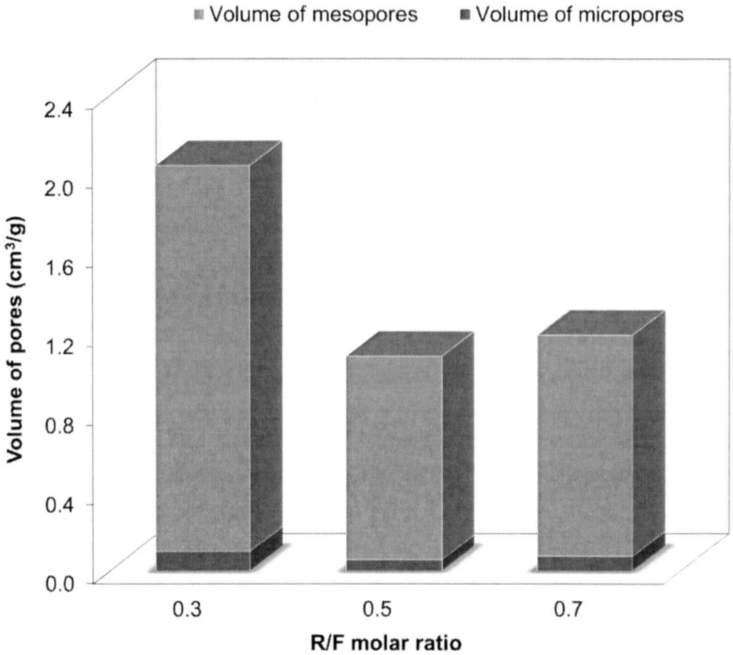

Figure 14. Evolution of the volume of mesopores of organic RF xerogels with a decrease in the R/F molar ratio from 0.5 (a) to 0.3 (b). Adapted from reference [23].

However, a recent published work reported that variations in the R/F molar ratio may also produce a significant change in the volume of micropores depending on the dilution ratio and pH values selected when preparing the precursor solutions. As pointed out before, the catalyst favors the addition reaction, which leads to the formation of a large number of small clusters interconnected by long necks. At the same time, a decrease in the dilution ratio results in an increase in the amount of resorcinol and formaldehyde used, which leads to a highly branched structure. Nonetheless, the results reported in this study demonstrated that the effect of these two chemical variables disappeared during the carbonization process, leaving the R/F molar ratio as the only significant chemical variable [55]. The authors postulated that an increase in the concentration of formaldehyde, i.e., a decrease in the R/F molar ratio, led to the formation of highly branched clusters. Consequently, carbon xerogels synthesized from precursor solutions with a higher concentration of formaldehyde displayed a greater number of highly branched clusters of larger size and so, larger micropore volumes were achieved. Furthermore, they also emphasized that there was the possibility that after the sol-gel reaction and curing steps, over-loaded and unreacted formaldehyde still remained among the clusters. The unreacted formaldehyde was eliminated during drying due to the low boiling point of formaldehyde. In this case, the place occupied by the formaldehyde molecules would become micropores resulting in higher micropore volumes.

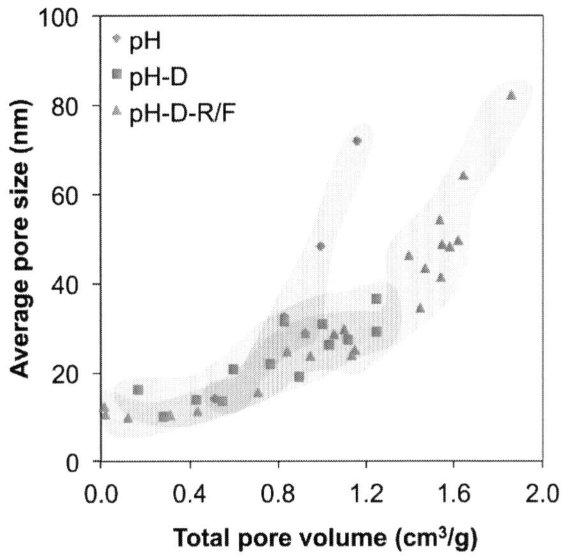

Figure 15. Improvement of the control of porosity of carbon xerogels achieved by the simultaneous modification of the molar ratio of resorcinol and formaldehyde, the pH and the dilution ratio. Reproduced by permission from reference [55].

There is no doubt that by modifying the R/F molar ratio, pH and D simultaneously it is possible to obtain a much wider range of materials than if only one variable is altered. Rey-Raap et al. reported the total pore volume versus the average pore size, as shown in Figure 15. In this figure three regions are highlighted. Each region indicates the materials synthesized from precursor solutions in which: i) different pH values were used while the dilution ratio and the R/F molar ratio were fixed at the most commonly used values in the literature, i.e., 5.7 and 0.5, respectively, ii) the pH and the dilution ratio were simultaneously modified while the R/F molar ratio was fixed at 0.5 and iii) the three chemical variables were simultaneously modified. As can be seen from Figure 15, the simultaneous adjustment of the R/F molar ratio, pH and dilution ratio yields average pore sizes within the range required for most applications where carbon gels are used (0- 200 nm), but with an increase in pore volume of 40% and also a much better control of the porous properties of the carbon xerogels.

Addition of Surfactants

The sol-gel reaction can be altered by adding additive agents such as surfactants, as these compounds can have a marked influence on the behavior of formaldehyde in the polymerization reaction [61, 62]. The most widely used classification for surfactants is based on the mode of dissociation of its molecules in solution. Accordingly, three groups can be distinguished: cationic, anionic and non-ionic surfactants.

The behavior of ionic surfactants (cationic and anionic) in the synthesis of RF gels has been studied by several research groups. Of all the possible cation surfactants, probably the most commonly used for synthesizing RF gels is hexadecyltrimethylammonium bromide (CTAB) [64, 76, 92, 93], whereas as anionic surfactant sodium derivatives, such as sodium docedilsulfate (SDS) [61] or sodium dodecylbenzenesulfonate (SDBS) [62, 63, 76, 94] are used. These surfactants form micelles in solution leaving their hydrophilic heads in contact with the precursor solution of the gel. The hydrophilic heads of cation surfactants are positively charged while the anionic surfactants have negative hydrophilic heads.

According to the polymeric reaction between resorcinol and formaldehyde, Matos et al. schematically show the effect of the cationic surfactant on the sol-gel reaction [61]. They postulate that the positive heads of the micelles interact electrostatically with resorcinol anions formed during the addition reaction, so that the addition of formaldehyde takes place close to the micellar surface, as shown in Figure 16.

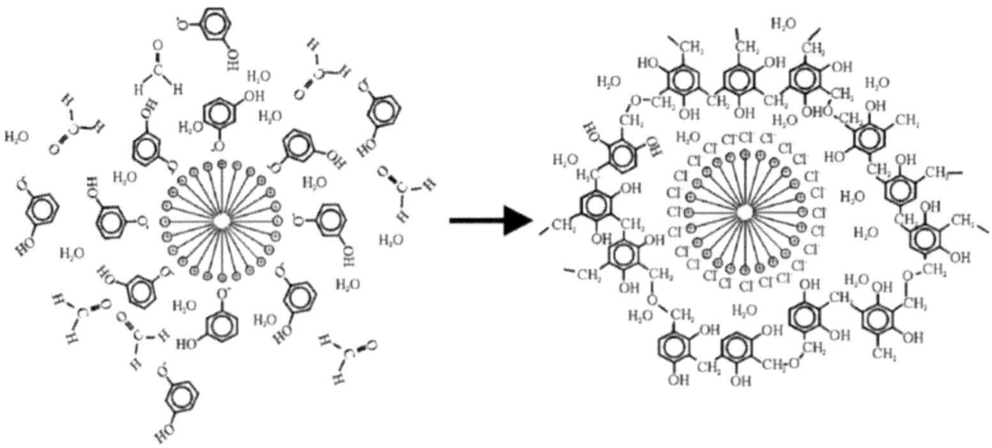

Figure 16. Scheme of the effect of the cationic surfactant on the polymeric structure of RF xerogels. Reproduced by permission from reference [61].

The effect of the cationic surfactant on the sol-gel reaction proposed by Matos et al. is in good agreement with the results of other authors [64]. Wu et al. showed that the use of cationic surfactant results in the formation of smaller clusters. Due to the effect of the surfactant, resorcinol anions are closer together, causing the addition of formaldehyde to proceed faster and, therefore, small clusters are formed. Furthermore, these authors indicated that an increase in the concentration of surfactant leads to greater shrinkage during drying, probably due to the higher surface tensions generated by small clusters. Accordingly, samples synthesized with a large amount of surfactant display a denser structure, as can be appreciated in Figure 17 which shows SEM images of samples synthesized with 5 wt. % and 8 wt. % of cationic surfactant, respectively [64]. In short, the use of a cationic surfactant results in less-mesoporous and more highly microporous materials. These porous properties make these materials suitable candidates for use as adsorbents, microporous membranes, porous electrolytes and catalyst supports [95].

On the other hand, the negative heads of anionic surfactants create electrostatic repulsion between resorcinol anions and the micellar surface, so that the polymeric structure is formed far away from the micelle, as shown in Figure 18 [61].

The repulsive forces between the resorcinol anions and surfactant cause the alienation of the resorcinol anions from the micelles. The addition of formaldehyde proceeds more slowly due to the large distances between the resorcinol anions, resulting in a structure with larger clusters, as shown in Figure 19 [63]. It can be seen from this figure that the use of anionic surfactant in the synthesis of RF gels leads to less compact structures with larger pore sizes.

Figure 17. SEM images of organic RF gels synthesized from RF precursor solutions with 5 wt. % (a) and 8 wt. % (b) of cationic surfactant. Reproduced with permission from reference [64].

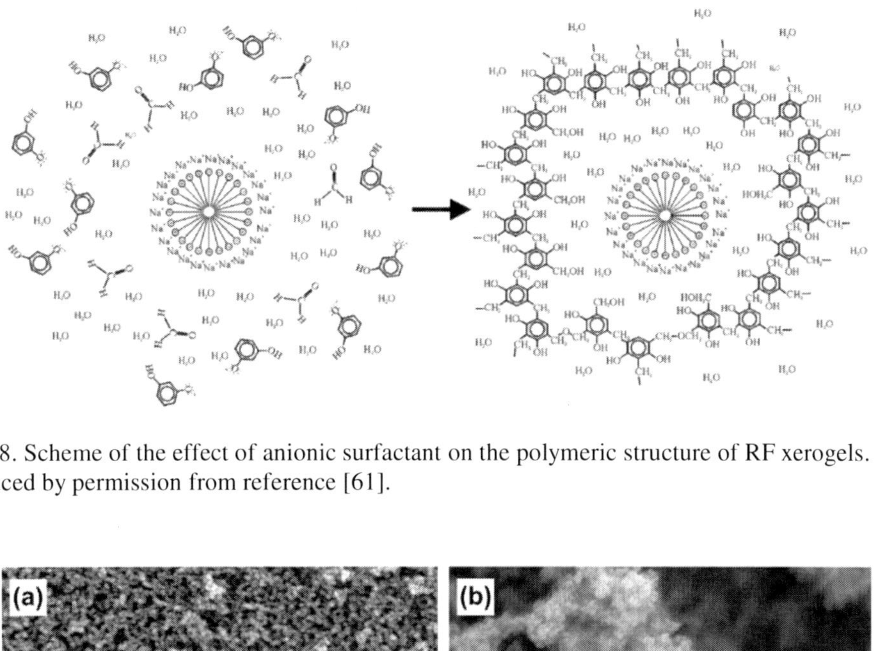

Figure 18. Scheme of the effect of anionic surfactant on the polymeric structure of RF xerogels. Reproduced by permission from reference [61].

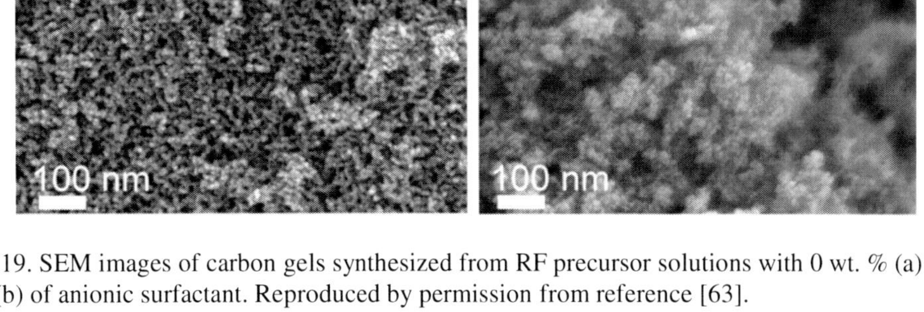

Figure 19. SEM images of carbon gels synthesized from RF precursor solutions with 0 wt. % (a) and 4 wt. % (b) of anionic surfactant. Reproduced by permission from reference [63].

The behavior of non-ionic surfactants in the synthesis of RF gels has also been widely studied. The commercial surfactants known as Span and Pluronic are the non-ionic surfactants most frequently found in the literature [54, 65, 76, 96]. However, it is worth noting that the effect that non-ionic surfactants have on the formation of the polymer structure is somewhat controversial due to differences in the method of preparation of the precursor solution used in each study. The most common method of preparation is based on emulsions [54, 61, 65, 76, 96]. First, the resorcinol-formaldehyde precursor solution is prepared. After gelation has taken place the surfactant is added. This step is followed by the curing and drying stages. Most of these studies indicate that the addition of a non-ionic surfactant has no significant effect on the sol-gel reaction between resorcinol and formaldehyde and therefore, does not alter the final properties of RF gels [61]. However, research work involving the dissolution of the surfactant in a solution of resorcinol prior to the addition of formaldehyde can also be found in the literature. Formaldehyde is then added to the resulting mixture and finally the sol-gel reaction, curing and drying stages are carried out [62, 95]. By using this method of preparation, Jirglovà et al. found that during the sol-gel reaction Span 80 may be linked to the RF structure through aliphatic C=C bonds or OH groups [62]. Consequently, they obtained more polymerized structures, resulting in more compact materials with low pore sizes. The properties of such materials make them highly suitable for use as molecular sieves.

Methanol As a Stabilizer of Formaldehyde

Formaldehyde is commercialized in solution due to its high propensity to polymerize. This propensity is the reason why formaldehyde, commonly used for the synthesis of RF gels (37 wt. % in water), is stabilized by methanol. The role of the methanol is to prevent the polymerization of formaldehyde by forming hemiacetal groups. The oxygen of the carbonyl group from formaldehyde is protonated under acidic conditions, yielding the formation of hemiformal, as shown in reaction A in Figure 20 [35]. The hemiformal formed is unstable and the equilibrium of the reaction shifts to favor the reactant. However, a sufficient amount of methanol allows the reaction to continue and form hemiacetals, as shown in the reaction B of Figure 20, which are, unlike hemiformal, more stable compounds [35]. Therefore, an increase in the concentration of methanol results in the formation of more hemiacetals, leaving less formaldehyde available to participate in the sol-gel reaction.

The concentration of methanol in the solution depends largely on the manufacturer. Table 1 shows concentrations of methanol ranging from 0.6 and 15 wt. % found in solutions of formaldehyde provided by different manufacturer.

Although there are a large number of studies in the literature that deal with the synthesis of RF gels (in which formaldehyde solutions stabilized with methanol are used), very few focus on the stabilizing effect of methanol. On the basis of the reaction mechanism shown in Figure 20, it can be said that the acidity of the medium clearly affects the stabilizing function of methanol. As seen throughout this chapter, the acidity of the medium is determined by the concentration of the catalyst. Low concentrations of catalyst (i.e., low pH values) not only result in the formation of fewer resorcinol anions but also promote the formation of hemiacetals. Therefore, not only are fewer resorcinol anions formed but also a lower amount of formaldehyde is available, leading to the formation of smaller clusters [33, 34]. Polymeric structures with small clusters give rise to small voids between clusters and; hence, samples which are macroporous at low pH values become mesoporous materials when the amount of methanol is increased, as shown in Figure 21.

Accordingly, materials with different porous properties can be obtained by using formaldehyde solutions provided by different manufacturers. However, the effects of the addition of methanol can be offset by an appropriate adjustment of other chemical variables such as the molar ratio between resorcinol and formaldehyde, the dilution ratio or the pH value of the precursor solution.

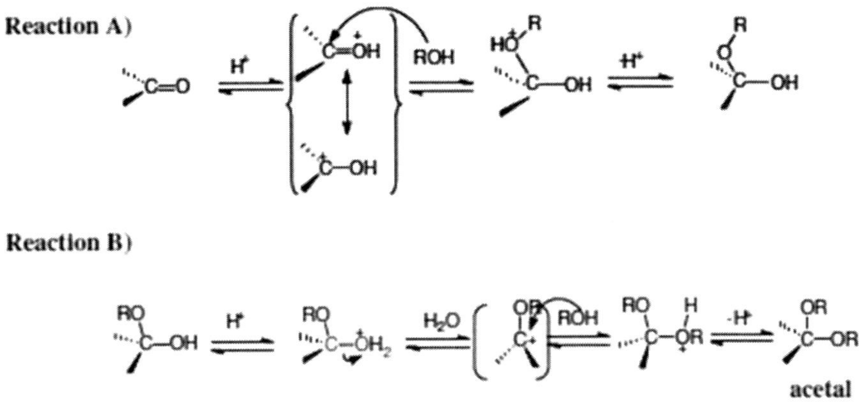

Figure 20. Stabilization mechanism of formaldehyde via the formation of hemiformal (a) and hemiacetals (b). Reproduced with permission from reference [35].

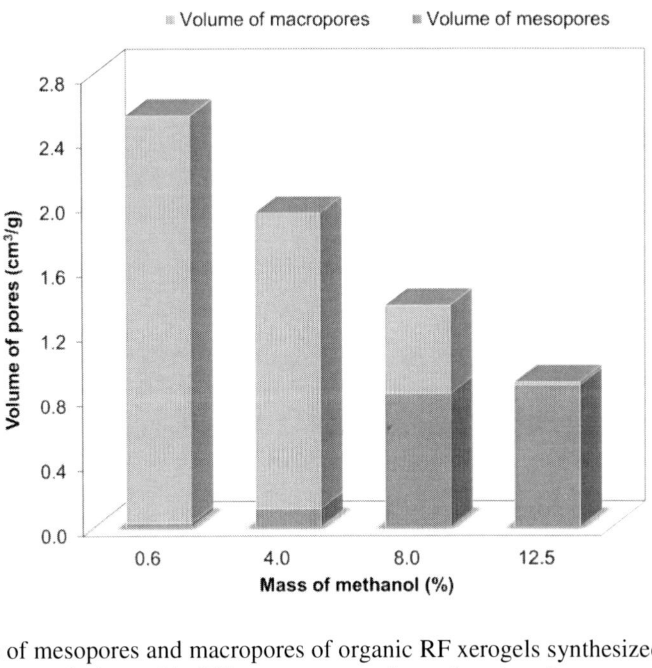

Figure 21. Volume of mesopores and macropores of organic RF xerogels synthesized by microwave heating from precursor solutions with different concentrations of methanol.

Table 1. Concentrations of methanol in solutions of formaldehyde provided by different manufacturers

Brand	Concentration of methanol (% wt.)
Fernández Rapado	≤ 1.3
Panreac	9 - 14
Solvech	9 - 15
Merck Chemicals	10
Oxidal	10 - 15
Sigma-aldrich	10 - 15
Spi-Chem	Non specified

REFERENCES

[1] Aegerter MA, Leventis N, Koebel MM. Aerogels handbook: *Springer;* 2011.

[2] Pekala RW. Organic aerogels from the polycondensation of resorcinol with formaldehyde. *J Mater Sci* 1989; 24 (9): 3221-7.

[3] Lee Y, Yoon JS, Suh DJ, Lee CH, Suh YW. 5-hydroxymethylfurfural as a potential monomer for the preparation of carbon aerogel. *Mater Chem Phys* 2012; 136 (2-3): 837-44.

[4] Scherdel C, Gayer R, Reichenauer G. Porous organic and carbon xerogels derived from alkaline aqueous phenol-formaldehyde solutions. *J Porous Mater* 2012; 19 (3): 351-60.

[5] Lin YF, Chen JL. Magnetic mesoporous Fe/carbon aerogel structures with enhanced arsenic removal efficiency. *J Colloid Interface Sci* 2014; 420: 74-9.

[6] Chandra S, Bag S, Bhar R, Pramanik P. Effect of transition and non-transition metals during the synthesis of carbon xerogels. *Microporous Mesoporous Mater.* 2011; 138 (1-3): 149-56.

[7] Job N, Pirard R, Marien J, Pirard JP. Porous carbon xerogels with texture tailored by pH control during sol-gel process. *Carbon* 2004; 42 (3): 619-28.

[8] Saliger R, Bock V, Petricevic R, Tillotson T, Geis S, Fricke J. Carbon aerogels from dilute catalysis of resorcinol with formaldehyde. *J Non-Cryst Solids* 1997; 221 (2-3): 144-50.

[9] Al-Muhtaseb SA, Ritter JA. Preparation and properties of resorcinol-formaldehyde organic and carbon gels. *Adv Mater* 2003; 15 (2): 101-14.

[10] Moreno-Castilla C, Maldonado-Hódar FJ. Carbon aerogels for catalysis applications: An overview. *Carbon* 2005; 43 (3): 455-65.

[11] Girgis BS, El-Sherif IY, Attia AA, Fathy NA. Textural and adsorption characteristics of carbon xerogel adsorbents for removal of Cu (II) ions from aqueous solution. *J Non-Cryst Solids* 2012; 358 (4): 741-7.

[12] Elkhatat AM, Al-Muhtaseb SA. Advances in tailoring resorcinol-formaldehyde organic and carbon gels. *Adv Mater* 2011; 23 (26): 2887-903.

[13] Wu D, Fu R, Sun Z, Yu Z. Low-density organic and carbon aerogels from the sol-gel polymerization of phenol with formaldehyde. *J Non-Cryst Solids* 2005; 351 (10-11): 915-21.

[14] Mukai SR, Tamitsuji C, Nishihara H, Tamon H. Preparation of mesoporous carbon gels from an inexpensive combination of phenol and formaldehyde. *Carbon* 2005; 43 (12): 2628-30.

[15] Long D, Zhang J, Yang J, Hu Z, Cheng G, Liu X, et al. Chemical state of nitrogen in carbon aerogels issued from phenol–melamine–formaldehyde gels. *Carbon* 2008; 46 (9): 1259-62.

[16] Li WC, Lu AH, Guo SC. Control of mesoporous structure of aerogels derived from cresol-formaldehyde. *J Colloid Interface Sci* 2002; 254 (1): 153-7.

[17] Zhang R, Li W, Li K, Lu C, Zhan L, Ling L. Effect of concentration of reactants on porosity of hydrogels, organic and carbon aerogels. Microporous Mesoporous Mater. 2004; 72 (1-3): 167-73.

[18] Zhu Y, Hu H, Li WC, Zhang X. Cresol-formaldehyde based carbon aerogel as electrode material for electrochemical capacitor. *J Power Sources* 2006; 162 (1): 738-42.

[19] Grzyb B, Hildenbrand C, Berthon-Fabry S, Bégin D, Job N, Rigacci A, et al. Functionalisation and chemical characterisation of cellulose-derived carbon aerogels. *Carbon* 2010; 48 (8): 2297-307.

[20] Szczurek A, Amaral-Labat G, Fierro V, Pizzi A, Masson E, Celzard A. The use of tannin to prepare carbon gels. Part I: Carbon aerogels. *Carbon* 2011; 49 (8): 2773-84.

[21] Braghiroli FL, Fierro V, Izquierdo MT, Parmentier J, Pizzi A, Celzard A. Nitrogen-doped carbon materials produced from hydrothermally treated tannin. *Carbon* 2012; 50 (15): 5411-20.

[22] Amaral-Labat G, Grishechko LI, Fierro V, Kuznetsov BN, Pizzi A, Celzard A. Tannin-based xerogels with distinctive porous structures. *Biomass and Bioenergy* 2013; 56: 437-45.

[23] Tamon H, Ishizaka H, Mikami M, Okazaki M. Porous structure of organic and carbon aerogels synthesized by sol-gel polycondensation of resorcinol with formaldehyde. *Carbon* 1997; 35 (6): 791-6.

[24] Job N, Sabatier F, Pirard JP, Crine M, Léonard A. Towards the production of carbon xerogel monoliths by optimizing convective drying conditions. *Carbon* 2006; 44 (12): 2534-42.

[25] Pérez-Caballero F, Peikolainen AL, Uibu M, Kuusik R, Volobujeva O, Koel M. Preparation of carbon aerogels from 5-methylresorcinol-formaldehyde gels. *Microporous Mesoporous Mater*. 2008; 108 (1-3): 230-6.

[26] Wu D, Fu R, Zhang S, Dresselhaus MS, Dresselhaus G. Preparation of low-density carbon aerogels by ambient pressure drying. *Carbon* 2004; 42 (10): 2033-9.

[27] Wu D, Fu R, Zhang S, Dresselhaus MS, Dresselhaus G. The preparation of carbon aerogels based upon the gelation of resorcinol-furfural in isopropanol with organic base catalyst. *J Non-Cryst Solids* 2004; 336 (1): 26-31.

[28] Tian HY, Buckley CE, Paskevicius M, Wang SB. Carbon aerogels from acetic acid catalysed resorcinol–furfural using supercritical drying for hydrogen storage. *The Journal of Supercritical Fluids* 2011; 55 (3): 1115-7.

[29] García BB, Liu D, Sepehri S, Candelaria S, Beckham DM, Savage LW, et al. Hexamethylenetetramine multiple catalysis as a porosity and pore size modifier in carbon cryogels. *J Non-Cryst Solids* 2010; 356 (33-34): 1620-5.

[30] Kocklenberg R, Mathieu B, Blacher S, Pirard R, Pirard JP, Sobry R, et al. Texture control of freeze-dried resorcinol–formaldehyde gels. *J Non-Cryst Solids* 1998; 225 (0): 8-13.
[31] Calvo EG, Juárez-Pérez EJ, Menéndez JA, Arenillas A. Fast microwave-assisted synthesis of tailored mesoporous carbon xerogels. *J Colloid Interface Sci* 2011; 357 (2): 541-7.
[32] Berthon S, Barbieri O, Ehrburger-Dolle F, Geissler E, Achard P, Bley F, et al. DLS and SAXS investigations of organic gels and aerogels. *J Non-Cryst Solids* 2001; 285 (1–3): 154-61.
[33] Fairén-Jiménez D, Carrasco-Marín F, Moreno-Castilla C. Adsorption of benzene, toluene, and xylenes on monolithic carbon aerogels from dry air flows. *Langmuir* 2007; 23 (20): 10095-101.
[34] Kiciński W, Szala M, Nita M. Structurally tailored carbon xerogels produced through a sol-gel process in a water-methanol-inorganic salt solution. *J Sol-Gel Sci Technol* 2011; 58 (1): 102-13.
[35] Zubizarreta L, Arenillas A, Domínguez A, Menéndez JA, Pis JJ. Development of microporous carbon xerogels by controlling synthesis conditions. *J Non-Cryst Solids* 2008; 354 (10-11): 817-25.
[36] Qin G, Guo S. Preparation of RF organic aerogels and carbon aerogels by alcoholic sol–gel process. *Carbon* 2001; 39 (12): 1935-7.
[37] Horikawa T, Hayashi J, Muroyama K. Controllability of pore characteristics of resorcinol-formaldehyde carbon aerogel. *Carbon* 2004; 42 (8-9): 1625-33.
[38] Mulik S, Sotiriou-Leventis C, Leventis LN. Time-efficient acid-catalyzed synthesis of resorcinol-formaldehyde aerogels. *Chem Mater* 2007; 19 (25): 6138-44.
[39] Gao P, Wang A, Wang X, Zhang T. Synthesis of highly ordered ir-containing mesoporous carbon materials by organic-organic self-assembly. Chem Mater 2008; 20 (5): 1881-8.
[40] Merzbacher CI, Meier SR, Pierce JR, Korwin ML. Carbon aerogels as broadband non-reflective materials. *J Non-Cryst Solids* 2001; 285 (1–3): 210-5.
[41] Laskowski J, Milow B, Ratke L. Subcritically dried resorcinol–formaldehyde aerogels from a base–acid catalyzed synthesis route. *Microporous Mesoporous Mater*. 2014; 197 (0): 308-15.
[42] Brandt R, Petricevic R, Pröbstle H, Fricke J. Acetic Acid Catalyzed Carbon Aerogels. *J Porous Mater* 2003; 10 (3): 171-8.
[43] Morales-Torres S, Maldonado-Hódar FJ, Pérez-Cadenas AF, Carrasco-Marín F. Textural and mechanical characteristics of carbon aerogels synthesized by polymerization of resorcinol and formaldehyde using alkali carbonates as basification agents. Phys Chem Chem Phys 2010; 12 (35): 10365-72.
[44] Gaca KZ, Sefcik J. Mechanism and kinetics of nanostructure evolution during early stages of resorcinol–formaldehyde polymerisation. *J Colloid Interface Sci* 2013; 406 (0): 51-9.
[45] Alegre C, Calvillo L, Moliner R, González-Expósito JA, Guillén-Villafuerte O, Huerta MVM, et al. Pt and PtRu electrocatalysts supported on carbon xerogels for direct methanol fuel cells. J Power Sources 2011; 196 (9): 4226-35.

[46] Zubizarreta L, Arenillas A, Pirard J-P, Pis JJ, Job N. Tailoring the textural properties of activated carbon xerogels by chemical activation with KOH. *Microporous Mesoporous Mater.* 2008; 115 (3): 480-90.

[47] Tamon H, Ishizaka H. Influence of gelation temperature and catalysts on the mesoporous structure of resorcinol-formaldehyde aerogels. *J Colloid Interface Sci* 2000; 223 (2): 305-7.

[48] Rey-Raap N, Angel Menéndez J, Arenillas A. RF xerogels with tailored porosity over the entire nanoscale. Microporous Mesoporous Mater. 2014; 195 (0): 266-75.

[49] Moreno AH, Arenillas A, Calvo EG, Bermúdez JM, Menéndez JA. Carbonisation of resorcinol-formaldehyde organic xerogels: Effect of temperature, particle size and heating rate on the porosity of carbon xerogels. *J Anal Appl Pyrolysis* 2013; 100: 111-6.

[50] Machado BF, Gomes HT, Serp P, Kalck P, Figueiredo JL, Faria JL. Carbon xerogel supported noble metal catalysts for fine chemical applications. Catal Today 2010; 149 (3-4): 358-64.

[51] Calvo EG, Lufrano F, Staiti P, Brigandì A, Arenillas A, Menéndez JA. Optimizing the electrochemical performance of aqueous symmetric supercapacitors based on an activated carbon xerogel. *J Power Sources* 2013.

[52] Taylor SJ, Haw MD, Sefcik J, Fletcher AJ. Gelation mechanism of resorcinol-formaldehyde gels investigated by dynamic light scattering. *Langmuir* 2014; 30 (34): 10231-40.

[53] Zubizarreta L, Arenillas A, Menéndez JA, Pis JJ, Pirard JP, Job N. Microwave drying as an effective method to obtain porous carbon xerogels. *J Non-Cryst Solids* 2008; 354 (33): 4024-6.

[54] Sharma CS, Upadhyay DK, Sharma A. Controlling the morphology of resorcinol-formaldehyde-based carbon xerogels by sol concentration, shearing, and surfactants. *Ind Eng Chem Res* 2009; 48 (17): 8030-6.

[55] Rey-Raap N, Angel Menéndez J, Arenillas A. Simultaneous adjustment of the main chemical variables to fine-tune the porosity of carbon xerogels. *Carbon* 2014; 78 (0): 490-9.

[56] Maldonado-Hódar FJ, Ferro-García MA, Rivera-Utrilla J, Moreno-Castilla C. Synthesis and textural characteristics of organic aerogels, transition-metal-containing organic aerogels and their carbonized derivatives. *Carbon* 1999; 37 (8): 1199-205.

[57] Yamamoto T, Nishimura T, Suzuki T, Tamon H. Control of mesoporosity of carbon gels prepared by sol-gel polycondensation and freeze drying. *J Non-Cryst Solids* 2001; 288 (1-3): 46-55.

[58] Feng Y, Wang J, Ge L, Jiang B, Miao L, Tanemura M. Pore size controllable preparation for low density porous nano-carbon. *J Nanosci Nanotechnol* 2013; 13 (10): 7012-5.

[59] Job N, Théry A, Pirard R, Marien J, Kocon L, Rouzaud J-N, et al. Carbon aerogels, cryogels and xerogels: Influence of the drying method on the textural properties of porous carbon materials. *Carbon* 2005; 43 (12): 2481-94.

[60] Lin C, Ritter JA. Effect of synthesis PH on the structure of carbon xerogels. *Carbon* 1997; 35 (9): 1271-8.

[61] Matos I, Fernandes S, Guerreiro L, Barata S, Ramos AM, Vital J, et al. The effect of surfactants on the porosity of carbon xerogels. *Microporous Mesoporous Mater.* 2006; 92 (1-3): 38-46.

[62] Jirglová H, Maldonado-Hódar FJ. Chemical interactions of surface-active agents with growing resorcinol-formaldehyde gels. *Langmuir* 2010; 26 (20): 16103-9.

[63] Worsley MA, Satcher Jr JH, Baumann TF. Influence of sodium dodecylbenzene sulfonate on the structure and properties of carbon aerogels. *J Non-Cryst Solids* 2010; 356 (3): 172-4.

[64] Wu D, Fu R, Dresselhaus MS, Dresselhaus G. Fabrication and nano-structure control of carbon aerogels via a microemulsion-templated sol-gel polymerization method. *Carbon* 2006; 44 (4): 675-81.

[65] Xu J, Wang A, Zhang T. A two-step synthesis of ordered mesoporous resorcinol-formaldehyde polymer and carbon. *Carbon* 2012; 50 (5): 1807-16.

[66] Despetis F, Barral K, Kocon L, Phalippou J. Effect of aging on mechanical properties of resorcinol-formaldehyde gels. *J Sol Gel Sci Technol* 2000; 19 (1-3): 829-31.

[67] Tonanon N, Siyasukh A, Tanthapanichakoon W, Nishihara H, Mukai SR, Tamon H. Improvement of mesoporosity of carbon cryogels by ultrasonic irradiation. *Carbon* 2005; 43 (3): 525-31.

[68] Tonanon N, Wareenin Y, Siyasukh A, Tanthapanichakoon W, Nishihara H, Mukai SR, et al. Preparation of resorcinol formaldehyde (RF) carbon gels: Use of ultrasonic irradiation followed by microwave drying. *J Non-Cryst Solids* 2006; 352 (52-54): 5683-6.

[69] Kang KY, Hong SJ, Lee BI, Lee JS. Enhanced electrochemical capacitance of nitrogen-doped carbon gels synthesized by microwave-assisted polymerization of resorcinol and formaldehyde. *Electrochem Commun* 2008; 10 (7): 1105-8.

[70] Czakkel O, Marthi K, Geissler E, László K. Influence of drying on the morphology of resorcinol-formaldehyde-based carbon gels. *Microporous Mesoporous Mater.* 2005; 86 (1-3): 124-33.

[71] Fu R, Zheng B, Liu J, Weiss S, Ying JY, Dresselhaus MS, et al. Fabrication of activated carbon fibers/carbon aerogels composites by gelation and supercritical drying in isopropanol. J Mater Res 2003; 18 (12): 2765-73.

[72] Wiener M, Reichenauer G, Scherb T, Fricke J. Accelerating the synthesis of carbon aerogel precursors. J Non-Cryst Solids 2004; 350 (0): 126-30.

[73] Moreno AH, Arenillas A, Calvo EG, Bermúdez JM, Menéndez JA. Carbonisation of resorcinol–formaldehyde organic xerogels: Effect of temperature, particle size and heating rate on the porosity of carbon xerogels. *J Anal Appl Pyrolysis* 2013; 100 (0): 111-6.

[74] Tamon H, Ishizaka H, Araki T, Okazaki M. Control of mesoporous structure of organic and carbon aerogels. *Carbon* 1998; 36 (9): 1257-62.

[75] Lin C, Ritter JA. Carbonization and activation of sol–gel derived carbon xerogels. *Carbon* 2000; 38 (6): 849-61.

[76] Maldonado-Hódar FJ, Jirglová H, Pérez-Cadenas AF, Morales-Torres S. Chemical control of the characteristics of Mo-doped carbon xerogels by surfactant-mediated synthesis. *Carbon* 2013; 51 (0): 213-23.

[77] Fairén-Jiménez D, Carrasco-Marín F, Moreno-Castilla C. Porosity and surface area of monolithic carbon aerogels prepared using alkaline carbonates and organic acids as polymerization catalysts. *Carbon* 2006; 44 (11): 2301-7.

[78] Contreras MS, Páez CA, Zubizarreta L, Léonard A, Blacher S, Olivera-Fuentes CG, et al. A comparison of physical activation of carbon xerogels with carbon dioxide with chemical activation using hydroxides. *Carbon* 2010; 48 (11): 3157-68.
[79] Conceição FL, Carrott PJM, Ribeiro Carrott MML. New carbon materials with high porosity in the 1–7 nm range obtained by chemical activation with phosphoric acid of resorcinol–formaldehyde aerogels. *Carbon* 2009; 47 (7): 1874-7.
[80] Zubizarreta L, Arenillas A, Pis JJ. Carbon materials for H2 storage. Int J Hydrogen Energy 2009; 34 (10): 4575-81.
[81] Calvo EG, Lufrano F, Staiti P, Brigandì A, Arenillas A, Menéndez JA. Optimizing the electrochemical performance of aqueous symmetric supercapacitors based on an activated carbon xerogel. *J Power Sources* 2013; 241 (0): 776-82.
[82] Lufrano F, Staiti P, Calvo EG, Juárez-Pérez EJ, Menéndez JA, Arenillas A. Carbon xerogel and manganese oxide capacitive materials for advanced supercapacitors. *Int J Electrochem Sci* 2011; 6 (3): 596-612.
[83] Li J, Wang X, Wang Y, Huang Q, Dai C, Gamboa S, et al. Structure and electrochemical properties of carbon aerogels synthesized at ambient temperatures as supercapacitors. *J Non-Cryst Solids* 2008; 354 (1): 19-24.
[84] Halama A, Szubzda B, Pasciak G. Carbon aerogels as electrode material for electrical double layer supercapacitors - Synthesis and properties. *Electrochim Acta* 2010; 55 (25): 7501-5.
[85] Mirzaeian M, Hall PJ. Preparation of controlled porosity carbon aerogels for en*ergy storage in rechargeable lithium oxygen batteries.* Electrochim Acta 2009; 54 (28): 7444-51.
[86] Zhang H, Xu H, Zhao C. Synthesis of morphology-controlled carbon hollow particles by carbonization of resorcinol-formaldehyde precursor microspheres and applications in lithium-ion batteries. *Mater Chem Phys* 2012; 133 (1): 429-36.
[87] Job N, Heinrichs B, Ferauche F, Noville F, Marien J, Pirard J-P. Hydrodechlorination of 1,2-dichloroethane on Pd–Ag catalysts supported on tailored texture carbon xerogels. *Catal Today* 2005; 102–103 (0): 234-41.
[88] Yamamoto T, Yoshida T, Suzuki T, Mukai SR, Tamon H. Dynamic and Static Light Scattering Study on the Sol-Gel Transition of Resorcinol-Formaldehyde Aqueous Solution. *J Colloid Interface Sci* 2002; 245 (2): 391-6.
[89] Schwan M, Ratke L. Flexibilisation of resorcinol-formaldehyde aerogels. *Journal of Materials Chemistry A* 2013; 1 (43): 13462-8.
[90] Wang F, Yao LF, Shen J, Guan DY, Fang Z. The effect of different ratio in carbon aerogel on pore structure in ambient dry. *Adv Mater Research.* 2014; 450-3.
[91] Feng J, Zhang C. Shrinkage and pore structure in preparation of carbon aerogels. *J Sol-Gel Sci Technol* 2011; 59 (2): 371-81.
[92] Lee KT, Oh SM. Novel synthesis of porous carbons with tunable pore size by surfactant-templated sol-gel process and carbonisation. *Chem Commun.* 2002; (22): 2722-3.
[93] Bruno MM, Cotella NG, Miras MC, Koch T, Seidler S, Barbero C. Characterization of monolithic porous carbon prepared from resorcinol/formaldehyde gels with cationic surfactant. *Colloids Surf, A* 2010; 358 (1–3): 13-20.

[94] Haghgoo M, Yousefi AA, Mehr MJZ. Nano porous structure of resorcinol-formaldehyde xerogels and aerogels: Effect of sodium dodecylbenzene sulfonate. *Iran Polym J Eng Ed* 2012; 21 (4): 211-9.

[95] Nishiyama N, Zheng T, Yamane Y, Egashira Y, Ueyama K. Microporous carbons prepared from cationic surfactant–resorcinol/formaldehyde composites. *Carbon* 2005; 43 (2): 269-74.

[96] Tonanon N, Tanthapanichakoon W, Yamamoto T, Nishihara H, Mukai SR, Tamon H. Influence of surfactants on porous properties of carbon cryogels prepared by sol-gel polycondensation of resorcinol and formaldehyde. *Carbon* 2003; 41 (15): 2981-90.

In: Formaldehyde
Editor: Amy Patton
ISBN: 978-1-63482-412-5

Chapter 3

FORMALDEHYDE: EXPOSURE AND GENOTOXICITY ASSESSMENTS AND POTENTIAL HEALTH EFFECTS

C. Ladeira[*,1,2] *and S. Viegas*[1,3]
[1]Environment Health Research Group, Escola Superior de Tecnologia da Saúde de Lisboa, Av. D. João II, Lisboa, Portugal
[2]Genetic and Metabolism Research Group, Escola Superior de Tecnologia da Saúde de Lisboa, Av. D. João II, Lisboa, Portugal
[3]Centro de Investigação em Saúde Pública (CISP/ENSP/UNL), Lisboa, Portugal

ABSTRACT

Worldwide formaldehyde is manipulated with diverse usage properties, since industrial purposes to health laboratory objectives, representing the economic importance of this chemical agent. Therefore, many people are exposed to formaldehyde environmentally and/or occupationally. Considering the latter, there was recommended occupational exposure limits based on threshold mechanisms, limit values and indoor guidelines.

Formaldehyde is classified by the International Agency for Cancer Research (IARC) as carcinogenic to humans (group 1), since a wide range of epidemiological studies in occupational exposure settings have suggested possible links between the concentration and duration of exposure and elevated risks of nasopharyngeal cancer, and others cancers, and more recently, with leukemia. Although there are different classifications, such as U.S. EPA that classified formaldehyde as a B1 compound, probable human carcinogen under the conditions of unusually high or prolonged exposure, on basis of limited evidence in humans but with sufficient evidence in animals.

Formaldehyde genotoxicity is well-known, being a direct-acting genotoxic compound positively associated for almost all genetic endpoints evaluated in bacteria, yeast, fungi, plants, insects, nematodes, and cultured mammalian cells. There are many human biomonitoring studies that associate formaldehyde occupational exposure to genomic instability, and consequently possible health effects. Besides the link with cancer, also other pathologies and symptoms are associated with formaldehyde exposure, namely respiratory disorders such as asthma, and allergic contact dermatitis.

[*] Corresponding author: carina.ladeira@estesl.ipl.pt.

Nowadays, there are efforts to reduce formaldehyde exposure, namely indoor. Europe and United States developed more strict regulation regarding formaldehyde emissions from materials containing this agent. Despite the regulations and restrictions, formaldehyde still continues to be difficult to eliminate or substitute, being biomonitoring an important tool to control possible future health effects.

1. INTRODUCTION

Formaldehyde (FA; CAS number: 50-00-0), also known as formalin, formol, and methyl aldehydes, is the most simple yet most reactive of all aldehydes, with the chemical formula CH2O. Its use and its physical, chemical and toxicological properties have been subject of a number of reviews (IARC, 1995; IPCS, 2004; IARC, 2006; Schulte, 2006;). It is a colorless gas at room temperature, flammable, and has a strong pungent smell (Liteplo et al., 2002; Pala et al., 2008; Zhang et al., 2009a). Aleksander Butlerov synthesized the chemical in 1859, but it was August Wilhelm von Hofmann who identified it in 1867, as the product formed from passing methanol and air over a heated platinum spiral. This method is still the basis for the industrial production of formaldehyde today, in which methanol is oxidized using a metal catalyst. By the early 20th century, the explosion of knowledge in chemistry and physics, coupled with demands for more innovative synthetic products, set the scene for the birth of new material plastics (Zhang et al., 2009a). It is a water-soluble gas with a pungent odor, is ubiquitous in the environment and naturally generated by processes that involve the combustion of organic materials, such as forest fires. Anthropogenic sources of formaldehyde can include combustion engines, aircraft, chemicals, textiles, timber, furniture, paints, paper and their manufacturing industries, as well as wood or charcoal fires, tobacco smoking, heating and cooking. Formaldehyde is also an endogenous substance that occurs as the result of various metabolic processes in most organisms, including in Humans (IARC, 2006).

Formaldehyde is an economically important chemical with an annual production of approximately 46 billion pounds worldwide. According to the Report on Carcinogens (NTP, 2011), formaldehyde ranks 25th in overall U.S. chemical production with more than 11 billion pounds produced each year. Formaldehyde production has increased steadily in China in recent years, with 7.5 million tons (16.5 billion pounds) of formaldehyde produced in 2007 (Zhang et al., 2009a, 2010; NTP, 2011).

Frequently used as a preservative or as a bactericide in medical laboratories and mortuaries, formaldehyde is also found in several and different products such as chemicals, particleboard, household products, glues, permanent press fabrics, paper product coatings, fiberboard, and plywood. It is also widely used as an industrial fungicide, germicide and disinfectant (IARC, 2006; NTP, 2011; OSHA, 2011). This last application is one of the most common and is commercially called as "formalin" and is used to describe a saturated solution of formaldehyde dissolved in water, typically with another agent, most commonly methanol, added to stabilize the solution. Formalin is typically 37% formaldehyde by weight (40% by volume) and 6-13% methanol by volume in water.

Formaldehyde has different formulations, such as casein formaldehyde, phenolic resins, urea formaldehyde, melamine formaldehyde and all have played an important role in the production of domestic and industrial goods that have become vital to everyday life. Casein formaldehyde (buttons, buckles, and knitting needles); phenolic resins (electrical and automobile insulations and other heavy industrial products, appliances toasters and radios); urea formaldehyde (picnic-ware, lampshades, varnishes, laminates and adhesives); and melanine formaldehyde (plastics more resistant to heat, water and detergents, cups, saucers and other domestic items) are formaldehyde-based materials that are broadly used (Zhang et al., 2009a; Kim et al., 2011; NTP, 2011). It is also widely used in molding compounds, glass wool and rock wool insulation, decorative laminates and textile treatments. Formaldehyde is now extensively used by industries across the globe. There are regulatory decisions regarding formaldehyde, such as occupational exposure limits (OELs) and drinking water standards, which have an economic impact that runs into millions of dollars (Zang et al., 2009).

Formaldehyde use in consumer products is very common, namely facial moisturizer, foundation, shower gels, shampoos, deodorants, hair conditioners, and hair styling gel and body lotions (Lefebvre et al., 2012). Additionally, and more recently, a novel hair cosmetic treatment called by Brazilian keratin treatment showed formaldehyde concentrations of 0.96% to 1.4%, well above to 0.2% the recommended for use in consumer products (Maneli et al., 2014).

There have been some restrictions to the use of formaldehyde in consumer products. For instance, in Canada, formaldehyde is a "controlled product" implying that the material safety data sheet for any commercial preparation containing more than 0.1% by mass of formaldehyde must mention its presence (Health Canada, 1999). With all these possible uses and applications is easy to understand that it is easily present in the air of outdoor and more commonly, in indoor environments.

2. EXPOSURE

Given its economic importance and widespread use, many people are exposed to formaldehyde environmentally and/or occupationally. Occupational exposure involves not only individuals employed in the direct manufacture of formaldehyde and products containing it, but also those in industries utilizing these products, such as construction (Liteplo *et al.*, 2002; Zhang *et al.*, 2009a), being reported frequently and is responsible for the higher exposures.

In 2010, Viegas and colleagues reported exposures in two different occupational settings: resins production and anatomy and pathology laboratories. In this study anatomy and pathology laboratories showed higher values for the two exposure metrics used namely, time-weighted average concentrations (TWA 8h) and ceiling concentrations. However, only in the case of ceiling concentrations the values were higher that the reference used (0.3 ppm). Previously, in a report prepared by Goyer et al. (2006) for L'Institut de Recherché Robert-Sauvé en Santé et en Sécurité du Travail (IRSST) found the difference between the two settings and, similar to the first reported research, they found that the handling of impregnated specimens was the task with higher exposure (>2 ppm).

Moreover, the health effects related with formaldehyde exposure, namely nasopharyngeal cancer, seems more related with the concentration than with the exposure time (IARC, 2006; Pyatt et al., 2008). Therefore, knowing the ceiling concentration is of extreme importance because allows identifying the exposure critical moments and more accurate and detail risk assessment (Viegas et al., 2009).

Table 1. Limit values of formaldehyde exposure by organization/country

Organization/Country	Limit Value (ppm)
NIOSH/United States	STEL – 0.1
ACGIH/United States	Ceiling – 0.3
OSHA/United States	TWA – 0.75 STEL – 2.0 Ceiling – 5.0
Canada	Ceiling - 0.3
INRS/France	TWA – 0.5 STEL — 1.0
NP 1796/Portugal	Ceiling – 0.3
Germany	TWA – 0.3 Ceiling — 1.0
Sweden	TWA - 0.5 Ceiling — 1.0

Legend: TWA = Time-weighted average exposure value; STEL = Short-term exposure value.
Sources: Agency for Toxic Substances and Disease Registry, 2007; Zhang et al.,2009.

The fact that some developed studies did not analyzed all information available and, also, the exposure metrics used are very different between studies (e.g., duration, peak exposure, cumulative exposure) may explain the results heterogeneity regarding exposure and health effects related in the studies developed until now (Bolt & Morfeld, 2013).

The formaldehyde occupational exposure limits of many countries are available on the International Labour Organization website and through the Registry of Toxic Effects of Chemical Substances database (RTECS#: LP8925000) maintained by National Institute for Occupational Safety and Health (NIOSH) (Zang et al., 2009a).

Regarding the reference limits for occupational exposure is important to notice that there are some differences between continents and even in the same continent, between countries (Table 1), and even the setting of diverse limits by different organizations, as in the United States.

More recently, in the 2nd International Formaldehyde Science Conference realized in 20132, Hermann Bolt explained the strategy of the Scientific Committee on Occupational Exposure Limits (SCOEL) of the European Union with respect to carcinogens in general (Bolt and Huici-Montagud, 2008) and to formaldehyde in particular (SCOEL, 2008). SCOEL has recommended an OEL for formaldehyde of 0.2 ppm, based on a carcinogenic threshold mechanism and the avoidance of any sensory irritation in exposed humans based on the studies available in 2008.

Regarding indoor environment, World Health Organization (2000) states as an indoor guideline value of 0.08 ppm (0.1 mg/m^3). This guideline as being consider preventive of carcinogenic effects in compliance with epidemiological findings. The main reasons are well described in the work published by Wolkoff and Nielsen (2010). Recent published work developed showed that formaldehyde concentrations have been continuously decreasing, down to current levels of 20 – 40 ppb (Bolt and Morfeld, 2013). Previously, a review developed by Salthammer (2011) of recent emission studies and indoor surveys carried out in industrialized countries has showed that the situation has improved significantly due to the efforts done in the last decades indicating that the average exposure of the population to formaldehyde seems to be generally below 0.05 ppm. The emission limitation of formaldehyde from building products and consumer goods has given an important contribution to this reduction.

3. Genotoxicity Assessment

The programme of the IARC "Monographs on the Evaluation of Carcinogenic Risk to Humans" represents a major example of hazard identification procedure (http://monographs.iarc.fr). For some genotoxic carcinogens the existence of a "practical" threshold is supported by studies on mechanisms and/or toxicokinetics. Such is the case of formaldehyde, for which a No-Observable Adverse Effect Limit (NOAEL) may be established from which a health-based exposure limit is derived (Bertazzi & Mutti, 2008).

The U.S. Environmental Protection Agency (EPA) classified formaldehyde as a B1 compound, probable human carcinogen under the conditions of unusually high or prolonged exposure, on basis of limited evidence in humans but with sufficient evidence in animals (US EPA, 2000).

Formaldehyde was long considered as a probable human carcinogen (Group 2A chemical) based on experimental animal studies and limited evidence of human carcinogenicity. Toxic responses induced by formaldehyde in the respiratory tract at high airborne concentrations, which include ulceration, hyperplasia, and squamous metaplasia, are "considered to contribute to the subsequent development of cancer" (IARC, 1995). However, the IARC reclassified formaldehyde as a human carcinogen (Group 1) in June 2004 based on "sufficient epidemiological evidence that formaldehyde causes nasopharyngeal cancer in humans" (IARC, 2006).

Formaldehyde is carcinogenic at the site of contact as a consequence of epithelial cell regenerative proliferation resulting from cytotoxicity and mutation (Li et al., 2007).

It has been estimated that as much as 22 – 42% of inhaled formaldehyde may be removed by mucus flow (Schlosser, 1999). Thus, the fundamental issue is the dose (Heck & Casanova, 2004). A minimum exposure concentration is required for formaldehyde to induce a toxic response; actually, formaldehyde is normally present in all cells and pathways for its metabolism already exists in all tissues and cells examined (Heck & Casanova, 2004). Formaldehyde is accepted as being toxic above certain doses and the chances of harmful effects are increased at room temperatures because of its volatility. When ingested, formaldehyde is rapidly metabolized and removed from the liver (Gulec et al., 2006). Formaldehyde is a direct-acting genotoxic compound and has been associated with positive results for almost all genetic endpoints evaluated in bacteria, yeast, fungi, plants, insects, nematodes, and cultured mammalian cells. It has caused base-pair gene mutations in *Salmonella typhimurium* and DNA adducts, DNA-protein crosslinks, DNA-DNA crosslinks, DNA single-strand breaks, unscheduled DNA synthesis, inhibition of DNA repair, gene mutations, cell transformation, and cytogenetics effects - chromosomal aberrations, sister chromatid exchanges, and micronuclei - in cultured mammalian cells (Suruda et al., 1993; Conaway et al., 1996; Heck & Casanova, 1999; Saito et al., 2005; Orsière et al., 2006; Speit et al., 2007; Zhang et al., 2009a; Jiang et al., 2010; Zhang et al., 2010; Kim et al., 2011; NTP, 2011). The mechanism by which formaldehyde causes cancer is not completely understood and most likely involves several modes of action. Formaldehyde exposure is associated with key events related to carcinogenicity, such as DNA reactivity, gene mutation, chromosomal breakage, aneuploidy, epigenetic effects, glutathione depletion, oxidative stress, and cytotoxicity-induced cellular proliferation (Saito *et al.*, et al., 2005; NTP, 2011). In summary, formaldehyde is genotoxic and induces both DNA damage and chromosome changes, frequently expressed as DNA-protein crosslinks. In recent years a number of reports indicate that formaldehyde can induce damage in circulating lymphocytes (Orsière et al., 2006; Zhang et al., 2009a).

It has been shown that a higher micronuclei frequency is directly associated with decreased efficiency of DNA repair and increased genome instability (Kirsch-Volders et al., 2006; Orsière et al., 2006).

The study from Ladeira et al. (2011) showed a statistically significant association between formaldehyde exposure and biomarkers of genotoxicity, namely micronuclei in lymphocytes and buccal cells, nucleoplasmic bridges, and nuclear buds. This can be explained in light of genomic instability, understood as an increased amount of mutations and/or chromosomal aberrations that cytogenetically translate into a greater frequency of changes in chromosome number and/or structure and in the formation of micronuclei (Zietkiewicz *et al.*, et al., 2009). Chromosome damage and effects upon lymphocytes arise because formaldehyde escapes from sites of direct contact, such as nose or mouth, originating nuclear alterations in the lymphocytes of those exposed (Goyer et al., 2004; Orsière et al., 2006; Zhang et al., 2009b; Speit et al., 2010). These results thus corroborate previous reports (He et al., 1998) that lymphocytes can be damaged by long term exposure to formaldehyde. Moreover, the changes in peripheral lymphocytes indicate that the cytogenetic effects triggered by formaldehyde can reach tissues faraway from the site of initial contact (Ye et al., 2005). Long term exposures to high concentrations of formaldehyde indeed appear to have a potential for DNA damage; these effects were well demonstrated in previous experimental studies with animals, where local genotoxic effects followed formaldehyde exposure, namely DNA-protein cross links and chromosome damage (IARC, 2006).

In humans, formaldehyde exposure is also associated with an increase in the frequency of micronuclei in buccal epithelial cells (Suruda et al., 1993; Burgaz et al., 2002; Speit et al., 2007; Ladeira et al. 2011). Suruda et al. (1993) claim that although changes in oral and nasal epithelial cells and peripheral blood cells do not indicate a direct mechanism leading to carcinogenesis, they present evidence that DNA alteration took place. It thus appears reasonable to conclude that formaldehyde is a cancer risk factor for those who are occupationally exposed in histopathology laboratories (IARC, 2006).

4. Health Effects Focus on Cancer

Formaldehyde irritates the nose and pharynx in humans and laboratory animals under a variety of circumstances (IARC, 2006) and also its inhalation is associated with respiratory symptoms, and eye, nose and throat irritation, dry skin, tearing eyes, conjunctivitis, sneezing, and coughing (Zhang et al., 2009a; Kim et al., 2011).

Nasal biopsies of workers chronically exposed to formaldehyde showed chronic inflammation, loss of cilia, mild dysplasia, hyperplasia and squamous metaplasia, although the latter finding has been inconsistent and may have been confounded by others exposures, such as to wood dust (IARC, 2006; 2012). The cytotoxicity of formaldehyde has been confirmed in numerous *in vitro* systems (IARC, 2006; Sul et al., 2007; IARC, 2006; IARC, 2012). Irritation of the nasal and upper respiratory tract is also noted in animal studies. Dose-dependent pathological findings include inflammation, hyperplasia, degenerative changes, necrosis and squamous metaplasia (IARC 2006; 2012). There is also some controversy regarding occupational asthma due to formaldehyde exposure. Asthma induced by inhaled formaldehyde may be classified as irritant induced asthma, as short exposures to high level formaldehyde are identified to cause a sudden onset of asthmatic symptoms called "Reactive airways dysfunction syndrome". Because of its airways-irritating properties, it may also aggravate pre-existing asthma (Kastner et al., 2011; Kim et al., 2011).

Formaldehyde is a known cause of allergic contact dermatitis. Skin sensitization to formaldehyde has been associated with many situations of dermal exposure, including with formaldehyde, formaldehyde-containing resins, formaldehyde-treated fabrics, formaldehyde containing household products, facial tissues, and others. Its exposure has been widely reported to cause dermal allergic reactions in occupationally exposed nurses, doctors, and dentists, as well as cosmetic workers, textile workers, and construction workers (Groot et al., 2009; Kim et al., 2011).

Formaldehyde exposure can cause a wide range of toxic effects such as formation of DNA-/protein cross-links, cytotoxicity, immune activation and sensory irritation (Gulec et al., 2006). Formaldehyde toxicity is thought to be mediated by the activation of free radical producing enzymes, and also by the inhibition of free radical scavenging systems, thereby enhancing the production of the reactive oxygen species (Gulec et al., 2006).

The sufficient evidence that formaldehyde is linked to cancer comes from a statistically significant excess of deaths from nasopharyngeal cancer in the largest and most informative cohort study of industrial workers by the U.S. National Cancer Institute (NCI), with a strong exposure-response correlation between the cancer mortality rate and peak cumulative exposures (Hauptmann et al., 2004, 2009). Collins & Lineker (2004) found an overall meta-relative risk for nasopharyngeal cancer also. After a thorough discussion of the epidemiologic, experimental and other relevant data, the IARC panel concluded that formaldehyde is a carcinogen in humans. However, it should be noted that some studies have argued that the IARC conclusion was premature and that the largest and most influential NCI study should be re-evaluated (Marsh & Youk, 2004; Marsh et al., 2010). Indeed, results in the literature are somewhat inconclusive. For example, Bosetti et al. (2008) concluded that workers and professionals exposed to formaldehyde showed no appreciable excess risk for oral and pharyngeal, sinonasal or lung cancers. However, for brain cancer and lymphohematopoietic neoplasms, there were modestly elevated risks in professionals, but not in industry workers. Walrath & Fraumeni (1983) showed a slightly elevated mortality by cancer in embalmers, but no excess of mortality from cancers of the respiratory tract including the nasal passages.

In their review, IARC also concluded that there was *"strong but not sufficient evidence for a causal association between leukemia and occupational exposure to formaldehyde"*. Much of the evidence for an association between formaldehyde exposure and leukemia comes from epidemiology studies, of which there are primarily three types: case-control studies in the general population, proportionate mortality studies of professionals, (e.g., funeral industry workers and pathologists) and cohort studies of industrial workers (Pinkerton et al., 2004; Zhang et al., 2010). Zhang et al. (2009a,b) state that formaldehyde may induce leukemia by damaging hematopoietic stem/progenitor cells circulating in the peripheral blood or by damaging the primitive pluripotent stem cells present within the nasal turbinates and/or olfactory mucosa. Since the previous evaluation (IARC, 2006), the NCI cohort of industrial workers in the USA has been updated with an additional ten years of mortality data. As in the previous analysis of leukemia (Hall et al., 1991), the association was stronger for myeloid leukemia and situations of peak exposure (Pinkerton et al., 2004; Freeman et al., 2009; NTP, 2011).

Other studies have argued that it is biologically implausible for formaldehyde to cause leukemia, being the primary arguments: (i) it is unlikely to reach the bone marrow and cause toxicity due to its highly reactive nature; (ii) there is no evidence that it can damage the stem and progenitor cells, the target cells for leukemogenesis; (iii) there is no credible experimental animal model for formaldehyde-induced leukemia; (iv) absence of an increase of blood formaldehyde levels after inhalative exposure as formaldehyde converting enzymes such as formaldehyde dehydrogenase and aldehydes dehydrogenase are present in many human tissues and also in human erytrocytes (Collins & Lineker, 2004; Heck & Casanova, 2004; Franks, 2005; Speit & Schmid, 2006, 2006; Neuss et al., 2010; NTP, 2011; Rhomberg et al., 2011) and cannot be directly related to occupation (Hayes et al., 1990). Indeed, IARC itself concluded *"based on the data available at this time, it was not possible to identify a mechanism for the induction of myeloid leukemia in humans"* and stated that *"this is an area needing more research"* (IARC, 2006). The last evaluation performed by IARC, concluded that there is *"sufficient evidence in humans for the carcinogenicity of formaldehyde. Formaldehyde causes cancer of the nasopharynx and leukemia. Also, a positive association has been observed between exposure to formaldehyde and sinonasal cancer"* (Baan et al., 2009; IARC, 2012).

5. Exposure Prevention and Regulatory Issues

Nowadays, in Europe, and in the scope of Registration, Evaluation, Authorization and Restriction of Chemicals (REACH) formaldehyde is being evaluate based on several aspects such as, the carcinogenic effect and the wide dispersive use leading to workers exposure.

Also in Europe there is actually strict regulation regarding formaldehyde emissions from furniture and other materials and where formaldehyde-containing materials, particularly resins, have been added to the product as a part of the production process, the product shall be tested and classified and, in some cases, prohibited to use indoors.

In 2010, the United States took some important measures also to reduce exposure in indoor environment: it was signed the Formaldehyde Standards for Composite Wood Products Act into law. This legislation, which adds a Title VI to the Toxic Substances Control Act (TSCA), establishes limits for formaldehyde emissions from composite wood products: hardwood plywood, medium-density fiberboard, and particleboard.

Despite all these efforts to regulate and restrain the contact to this chemical agent the substitution in some settings is difficult and can take a really long time until the use is complete eliminate and the risk reduced. Since 1987, there have been some efforts to substitute formaldehyde in some of the activities developed in the anatomy and pathology laboratories (Titford, 2005). However, the new products also present some toxicity and this implies also precautionary procedures (Titford & Horenstein, 2005). Besides this, the fixation quality is difficult to equalize the one obtain with formaldehyde solution and there is no study until now that present a good relation between cost and advantages of the new products leading to low acceptation of the professionals (Prentö & Lyon, 1997; Goyer et al., 2004; Jones, 2007).

REFERENCES

Baan, R., Grosse, Y., Straif, K., Secretan, B., El Ghissassi, F., Bouvard, V., Benbrahim-Tallaa, L., Guha, N., Freeman, C., Galichet, L. & Cogliano, V. (2009). A review of human carcinogens—Part F: Chemical agents and related occupations. *The Lancet Oncology*. 10 (12). p.pp. 1143–1144.

Bertazzi, P.A. & Mutti, A. (2008). Biomarkers, Disease Mechanisms and their Role in Regulatory Decisions. In: C. Wild, P. Vineis, & S. Garte (eds.). *Molecular Epidemiology of Chronic Diseases*. [Online]. John Wiley & Sons, Ltd, pp. 243–254. Available from: http://onlinelibrary.wiley. com/doi/10.1002 /9780470725726.ch19/summary.

Bolt, H.M. & Huici-Montagud, A. (2008). Strategy of the Scientific Committee on Occupational Exposure Limits (SCOEL) in the derivation of occupational exposure limits for carcinogens and mutagens. *Arch Toxicol* 82(1):61–64.

Bolt, H.M. & Morfeld, P. (2013). New results on formaldehyde: the 2nd International Formaldehyde Science Conference (Madrid, 19–20 April 2012). *Arch Toxicol* 87:217–222.

Bosetti, C., McLaughlin, J.K., Tarone, R.E., Pira, E. & Vecchia, C.L. (2008). Formaldehyde and cancer risk: a quantitative review of cohort studies through 2006. *Annals of Oncology*. 19 (1). p.pp. 29–43.

Burgaz, S., Erdem, O., Cakmak, G., Erdem, N., Karakaya, A. & Karakaya, A.E. (2002). Cytogenetic analysis of buccal cells from shoe-workers and pathology and anatomy laboratory workers exposed to n-hexane, toluene, methyl ethyl ketone and formaldehyde. *Biomarkers: biochemical indicators of exposure, response, and susceptibility to chemicals*. 7 (2). p.pp. 151–161.

Collins, J.J. & Lineker, G.A. (2004). A review and meta-analysis of formaldehyde exposure and leukemia. *Regulatory Toxicology and Pharmacology*. 40 (2). p.pp. 81–91.

Conaway, C.C., Whysner, J., Verna, L.K. & Williams, G.M. (1996). Formaldehyde mechanistic data and risk assessment: endogenous protection from DNA adduct formation. *Pharmacology & therapeutics*. 71 (1-2). p.pp. 29–55.

Franks, S.J. (2005). A mathematical model for the absorption and metabolism of formaldehyde vapour by humans. *Toxicology and Applied Pharmacology*. 206 (3). p.pp. 309–320.

Freeman, L.E.B., Blair, A., Lubin, J.H., Stewart, P.A., Hayes, R.B., Hoover, R.N. & Hauptmann, M. (2009). Mortality From Lymphohematopoietic Malignancies Among

Workers in Formaldehyde Industries: The National Cancer Institute Cohort. *Journal of the National Cancer Institute*. 101 (10). p.pp. 751–761.

Goyer, N., Beaudry, C., Bégin, D., Bouchard, M., Buissonet, S., Carrier, G., Gely, O., Gérin, M., Lavoué, L., Lefebvre, P., Noisel, N., Perraut, G. & Roberge, B. (2004). *Impacts d'un abaissement de la valeur d'exposition admissible au formaldéhyde : industries de fabrication de formaldéhde et de résines à base de formaldéhyde.*

Groot, A.C., Flyvholm, M.-A., Lensen, G., Menné, T. & Coenraads, P.-J. (2009). Formaldehyde-releasers: relationship to formaldehyde contact allergy. Contact allergy to formaldehyde and inventory of formaldehyde-releasers. *Contact dermatitis*. 61 (2). p.pp. 63–85.

Gulec, M., Gurel, A. & Armutcu, F. (2006). Vitamin E protects against oxidative damage caused by formaldehyde in the liver and plasma of rats. *Molecular and cellular biochemistry*. 290 (1-2). p.pp. 61–67.

Hall, A., Harrington, J.M. & Aw, T.C. (1991). Mortality study of British pathologists. *American journal of industrial medicine*. 20 (1). p.pp. 83–89.

Hauptmann, M., Lubin, J.H., Stewart, P.A., Hayes, R.B. & Blair, A. (2004). Mortality from solid cancers among workers in formaldehyde industries. *American journal of epidemiology*. 159 (12). p.pp. 1117–1130.

Hauptmann, M., Stewart, P.A., Lubin, J.H., Freeman, L.E.B., Hornung, R.W., Herrick, R.F., Hoover, R.N., Fraumeni, J.F., Blair, A. & Hayes, R.B. (2009). Mortality From Lymphohematopoietic Malignancies and Brain Cancer Among Embalmers Exposed to Formaldehyde. *Journal of the National Cancer Institute*. [Online]. Available from: http://jnci. oxfordjournals.org/content/early/2009/11/20/jnci.djp416.

Hayes, R.B., Blair, A., Stewart, P.A., Herrick, R.F. & Mahar, H. (1990). Mortality of U.S. embalmers and funeral directors. *American journal of industrial medicine*. 18 (6). p.pp. 641–652.

He, J.L., Jin, L.F. & Jin, H.Y. (1998). Detection of cytogenetic effects in peripheral lymphocytes of students exposed to formaldehyde with cytokinesis-blocked micronucleus assay. *Biomedical and environmental sciences: BES*. 11 (1). p.pp. 87–92.

Heck, H. & Casanova, M. (2004). The implausibility of leukemia induction by formaldehyde: a critical review of the biological evidence on distant-site toxicity. *Regulatory Toxicology and Pharmacology*. 40 (2). p.pp. 92–106.

Heck, H. & Casanova, M. (1999). Pharmacodynamics of formaldehyde: applications of a model for the arrest of DNA replication by DNA-protein cross-links. *Toxicology and applied pharmacology*. 160 (1). p.pp. 86–100.

IARC (2012). A Review of Human Carcinogens: Chemical Agents and Related Occupations. 100F.

IARC Working Group on the Evaluation of Carcinogenic Risks to Humans (2006). Formaldehyde, 2-butoxyethanol and 1-tert-butoxypropan-2-ol. IARC monographs on the evaluation of carcinogenic risks to humans / World Health Organization, International Agency for Research on Cancer. 88. p.pp. 1–478.

IARC (1995). IARC Monographs on the evaluation of carcinogenic risks to humans - wood dust and formaldehyde. 62.

Jiang, S., Yu, L., Cheng, J., Leng, S., Dai, Y., Zhang, Y., Niu, Y., Yan, H., Qu, W., Zhang, C., Zhang, K., Yang, R., Zhou, L. & Zheng, Y. (2010). Genomic damages in peripheral blood lymphocytes and association with polymorphisms of three glutathione Stransferases in workers exposed to formaldehyde. *Mutation Research/Genetic Toxicology and Environmental Mutagenesis*. 695 (1-2). p.pp. 9–15.

Jones, M. L. (2007). How formalin affects the outcome of routine and special stains. *Biotechnic and Histochemistry*. 82 (3) 155-159.

Kastner, P.E., Casset, A. & Pons, F. (2011). Formaldehyde interferes with airway epithelium integrity and functions in a dose- and time-dependent manner. *Toxicology letters*. 200 (1-2). p.pp. 109–116.

Kim, K.-H., Jahan, S.A. & Lee, J.-T. (2011). Exposure to Formaldehyde and Its Potential Human Health Hazards. *Journal of Environmental Science and Health, Part C*. 29 (4). p.pp. 277–299.

Kirsch-Volders, M., Mateuca, R.A., Roelants, M., Tremp, A., Zeiger, E., Bonassi, S., Holland, N., Chang, W.P., Aka, P.V., Deboeck, M., Godderis, L., Haufroid, V., Ishikawa, H., Laffon, B., Marcos, R., Migliore, L., Norppa, H., Teixeira, J.P., Zijno, A. & Fenech, M. (2006). The effects of GSTM1 and GSTT1 polymorphisms on micronucleus frequencies in human lymphocytes in vivo. Cancer epidemiology, biomarkers & prevention: a publication of the American Association for Cancer Research, cosponsored by the American Society of Preventive Oncology. 15 (5). p.pp. 1038–1042.

Ladeira, C., Viegas, S., Carolino, E., Prista, J., Gomes, M.C. & Brito, M. (2011). Genotoxicity biomarkers in occupational exposure to formaldehyde-the case of histopathology laboratories. *Mutation research*. 721 (1). p.pp. 15–20.

Li, G.-Y., Lee, H.-Y., Shin, H.-S., Kim, H.-Y., Lim, C.-H. & Lee, B.-H. (2007). Identification of Gene Markers for Formaldehyde Exposure in Humans. Environmental Health Perspectives. [Online]. Available from: http://www.ehponline.org/docs/2007/10180 /abstract.html.

Liteplo, R., World Health Organization, International Labour Organisation, United Nations Environment Programme & Inter-Organization Programme for the Sound Management of Chemicals (2002). *Formaldehyde*. Geneva: World Health Organization.

Maneli, M.H., Smith, P. & Khumalo, N.P. (2014). Elevated formaldehyde concentration in 'Brazilian keratin type' hair-straightening products: a cross-sectional study. *Journal of the American Academy of Dermatology*. 70 (2). p.pp. 276–280.

Marsh, G.M. & Youk, A.O. (2004). Reevaluation of mortality risks from leukemia in the formaldehyde cohort study of the National Cancer Institute. *Regulatory Toxicology and Pharmacology*. 40 (2). p.pp. 113–124.

Marsh, G.M., Youk, A.O., Morfeld, P., Collins, J.J. & Symons, J.M. (2010). Incomplete follow-up in the National Cancer Institute's formaldehyde worker study and the impact on subsequent reanalyses and causal evaluations. *Regulatory Toxicology and Pharmacology*. 58 (2). p.pp. 233–236.

Neuss, S., Moepps, B. & Speit, G. (2010). Exposure of human nasal epithelial cells to formaldehyde does not lead to DNA damage in lymphocytes after co-cultivation. *Mutagenesis*. 25 (4). p.pp. 359–364.

NTP (2011). *Report on Carcinogens, Twelfth Edition.*

Orsière, T., Sari-Minodier, I., Iarmarcovai, G. & Botta, A. (2006). Genotoxic risk assessment of pathology and anatomy laboratory workers exposed to formaldehyde by use of personal air sampling and analysis of DNA damage in peripheral lymphocytes. *Mutation Research/Genetic Toxicology and Environmental Mutagenesis*. 605 (1-2). p.pp. 30–41.

Pala, M., Ugolini, D., Ceppi, M., Rizzo, F., Maiorana, L., Bolognesi, C., Schilirò, T., Gilli, G., Bigatti, P., Bono, R. & Vecchio, D. (2008). Occupational exposure to formaldehyde and biological monitoring of Research Institute workers. *Cancer Detection and Prevention*. 32 (2). p.pp. 121–126.

Pinkerton, L.E., Hein, M.J. & Stayner, L.T. (2004). Mortality among a cohort of garment workers exposed to formaldehyde: an update. *Occupational and Environmental Medicine*. 61 (3). p.pp. 193–200.

Prentö, P., Luon, H. (1997) Commercial formalin substitutes for histopathology. *Biotechnic & Histochemistry*. 72 (5) 273-282.

Rhomberg, L.R., Bailey, L.A., Goodman, J.E., Hamade, A.K. & Mayfield, D. (2011). Is exposure to formaldehyde in air causally associated with leukemia?—A hypothesis-based weight-of-evidence analysis. *Critical Reviews in Toxicology*. 41 (7). p.pp. 555–621.

Saito, Y., Nishio, K., Yoshida, Y. & Niki, E. (2005). Cytotoxic effect of formaldehyde with free radicals via increment of cellular reactive oxygen species. *Toxicology*. 210 (2-3). p.pp. 235–245.

Salthammer, T. (2011). Evaluation of Formaldehyde Guideline Values for Indoor Air. *Epidemiology*. 22 (1). p.pp. S39.

Schlosser, P.M. (1999). Relative roles of convection and chemical reaction for the disposition of formaldehyde and ozone in nasal mucus. *Inhalation toxicology*. 11 (10). p.pp. 967–980.

Schulte, A. (2006). Assessment of the carcinogenicity of Formaldehyde (CAS No. 50-00-5): Bericht zur Bewertung der Karzinogenität von Formaldehyd. Berlin: BfR, Pressestelle.

Speit, G., Gelbke, H.-P., Pallapies, D. & Morfeld, P. (2010). Occupational Exposure to Formaldehyde, Hematotoxicity and Leukemia-Specific Chromosome Changes in Cultured Myeloid Progenitor Cells - Letter. *Cancer Epidemiology Biomarkers & Prevention*. 19 (7). p.pp. 1882–1884.

Speit, G., Schmid, O., Fröhler-Keller, M., Lang, I. & Triebig, G. (2007). Assessment of local genotoxic effects of formaldehyde in humans measured by the micronucleus test with exfoliated buccal mucosa cells. *Mutation Research/Genetic Toxicology and Environmental Mutagenesis*. 627 (2). p.pp. 129–135.

Speit, G. & Schmid, O. (2006). Local genotoxic effects of formaldehyde in humans measured by the micronucleus test with exfoliated epithelial cells. *Mutation Research/Reviews in Mutation Research*. 613 (1). p.pp. 1–9.

Sul, D., Kim, H., Oh, E., Phark, S., Cho, E., Choi, S., Kang, H.-S., Kim, E.-M., Hwang, K.-W. & Jung, W.-W. (2007). Gene expression profiling in lung tissues from rats exposed to formaldehyde. *Archives of Toxicology*. 81 (8). p.pp. 589–597.

Suruda, A., Schulte, P., Boeniger, M., Hayes, R.B., Livingston, G.K., Steenland, K., Stewart, P., Herrick, R., Douthit, D. & Fingerhut, M.A. (1993). Cytogenetic effects of formaldehyde exposure in students of mortuary science. Cancer epidemiology, biomarkers & prevention: a publication of the American Association for Cancer Research, cosponsored by the American Society of Preventive Oncology. 2 (5). p.pp. 453–460.

Titford, M. (2005). The long history of hematoxylin. Biotechnic and Histochemistry. 80 (2) 73-78.

Titford, M. E. & Horenstein, M. G. (2005). Histomorphologic assessment of formalin substitute fixatives diagnostic surgical pathology. *Archives of Pathology and Laboratory Medicine*. 129 (4) 502-506.

US EPA, O. (2000). Formaldehyde | Technology Transfer Network Air Toxics Web site | US EPA. [Online]. Available from: http://www.epa.gov/ttn /atw/hlthef/formalde.html.

Viegas, S., Ladeira, C., Nunes, C., Malta-Vacas, J., Gomes, M., Brito, M., Mendonca, P. & Prista, J. (2010). Genotoxic effects in occupational exposure to formaldehyde: A study in anatomy and pathology laboratories and formaldehyde-resins production. *Journal of occupational medicine and toxicology (London, England)*. 5 (1). p.p. 25.

Walrath, J. & Fraumeni, J.F., Jr. (1983). Mortality patterns among embalmers. *International journal of cancer. Journal international du cancer*. 31 (4). p.pp. 407–411.

WHO. (2000). Air quality guidelines for Europe, 2nd ed. World Health Organization, Regional Office for Europe, Copenhagen, pp 87–91.

Wolkoff, P. & Nielsen, G. (2010). Non-cancer effects of formaldehyde and relevance for indoor air guideline. *Environ Intern* 36:788–799.

Ye, X., Yan, W., Xie, H., Zhao, M. & Ying, C. (2005). Cytogenetic analysis of nasal mucosa cells and lymphocytes from high-level long-term formaldehyde exposed workers and low-level short-term exposed waiters. *Mutation Research*. 588 (1). p.pp. 22–27

Zhang, L., Freeman, L.E.B., Nakamura, J., Hecht, S.S., Vandenberg, J.J., Smith, M.T. & Sonawane, B.R. (2009a). Formaldehyde and leukemia: Epidemiology, potential mechanisms, and implications for risk assessment. *Environmental and Molecular Mutagenesis*. p.p. NA–NA.

Zhang, L., Steinmaus, C., Eastmond, D.A., Xin, X.K. & Smith, M.T. (2009b). Formaldehyde exposure and leukemia: A new meta-analysis and potential mechanisms. *Mutation Research/Reviews in Mutation Research*. 681 (2-3). p.pp. 150–168.

Zhang, L., Tang, X., Rothman, N., Vermeulen, R., Ji, Z., Shen, M., Qiu, C., Guo, W., Liu, S., Reiss, B., Freeman, L.B., Ge, Y., Hubbard, A.E., Hua, M., Blair, A., Galvan, N., Ruan, X., Alter, B.P., Xin, K.X., Li, S., Moore, L.E., Kim, S., Xie, Y., Hayes, R.B., Azuma, M., Hauptmann, M., Xiong, J., Stewart, P., Li, L., Rappaport, S.M., Huang, H., Fraumeni, J.F., Smith, M.T. & Lan, Q. (2010). Occupational Exposure to Formaldehyde, Hematotoxicity, and Leukemia-Specific Chromosome Changes in Cultured Myeloid Progenitor Cells. *Cancer Epidemiology Biomarkers & Prevention*. 19 (1). p.pp. 80–88.

Zietkiewicz, E., Wojda, A. & Witt, M. (2009). Cytogenetic perspective of ageing and longevity in men and women. *Journal of Applied Genetics*. 50 (3). p.pp. 261–273.

In: Formaldehyde
Editor: Amy Patton
ISBN: 978-1-63482-412-5

Chapter 4

FORMALDEHYDE AND ALZHEIMER'S DISEASE: A BRIEF HISTORY

Joshua D. Rizak
Key State Laboratory of Animal Models and Disease Mechanisms, Kunming Institute of Zoology, Chinese Academy of Sciences, Kunming, Yunnan, P. R. China

ABSTRACT

Formaldehyde has recently been placed at the forefront of Alzheimer's disease research. The reasons for this impetus is multi-fold and because the properties of formaldehyde, a planar polar single carbon aldehyde, have strong relevance to the disease. Alzheimer's disease is an age-related neurodegenerative dementia that presents with cognitive decline and memory impairments and is diagnosed with neuritic plaques, neurofibrillary tangles and neuronal atrophy. First, formaldehyde has been found to be an age-related factor whose concentrations in blood and cerebral spinal fluid increase with age and has been found to be higher in Alzheimer's patients post-mortem. Formaldehyde and methanol, a one-carbon precursor of formaldehyde, have been found to cause memory impairments in laboratory animals. Moreover, formaldehyde is a known protein cross-linker that has been shown to cause the aggregation of both amyloid peptides and tau protein as well as tau protein phosphorylation *in vitro*. These properties are central to the two main pathological hallmarks of Alzheimer's disease: the amyloid plaque and neurofibrillary tangle. As such, chronic methanol exposure has been found to induce all of these hallmarks, including memory impairments, in rhesus monkeys. While these findings have implications related to exogenous exposure to both methanol and formaldehyde, there are a number of indications that endogenous metabolic processes may also contribute formaldehyde to the pathological state. Most notably is the condition of hyperhomocysteremia in Alzheimer's disease. Hyperhomocysteremia represents an imbalance of the one-carbon metabolic cycle that favors the metabolite homocysteine, while potentially contributing free one-carbon moieties like formaldehyde to the blood or cerebral spinal fluid. However, no correlation studies between these two related metabolites and Alzheimer's disease have been investigated to date. Nonetheless, formaldehyde offers a unique chemical toxicity that has the potential to explain a large gambit of the pathological progression of the disease; from age-related progression, memory impairments and genetic risk factors. Although there is considerable implications of formaldehyde, such as in the development of novel therapies for the treatment of Alzheimer's disease, very little research and focus have been given to one-carbon toxicity in the past. This brief review serves to focus related research and raise awareness of formaldehyde in Alzheimer's disease pathology so that it may receive better consideration in the future.

The relationship between formaldehyde, a single carbon aldehyde, and Alzheimer's disease, the world's leading form of dementia, has an interesting history; albeit one that has been regularly overlooked until recently. In 2010, Dr. RongQiao He and colleagues at the Chinese Academy of Sciences discovered that formaldehyde levels are elevated in the urine and blood of aging populations, and that these elevations are more prominent in patients with Alzheimer's Disease (Wang et al., 2010, Tong et al., 2011 & 2013). Moreover, they found evidence in post-mortem brain tissue of Alzheimer's patients that formaldehyde accumulated in the hippocampus (Tong et al., 2013), a brain structure of the temporal lobe with a well-documented role in memory consolidation and retrieval (reviewed in Preston & Eichenbaum, 2013; Stuchlik, 2014). Interestingly, formaldehyde levels in urine were found to be inversely correlated to cognitive abilities in dementia patients (Tong et al., 2011). These evidences have brought formaldehyde to the forefront of Alzheimer's disease research, most notably, because Alzheimer's disease is an age related and progressive neurodegenerative dementia that is associated with cognitive decline and memory impairments.

Even before this connection was firmly established, formaldehyde, itself, had undergone extensive study into its toxic effects (reviewed in He et al., 2010) and was found to have properties similar to pathological markers of Alzheimer's disease. Two notable examples include (1) memory and cognitive impairments in laboratory animals (Malak et al., 2003; Zararsiz et al., 2007; Lu et al., 2008) and in workplace associated studies in human beings (Kilburn et al., 1987) and (2) protein cross-linking effects due to the chemical nature of formaldehyde (Bogdanffy et al., 1987; Toews et al., 2008). This second property, despite being exploited for molecular biology studies to identify interacting proteins (Orlando, 2000), has been relatively overlooked with regards to pathological effects until recently. In the case of Alzheimer's disease pathology, the relevance of this chemical property of formaldehyde should not be understated given that the two most common pathological hallmarks of the disease are extracellular aggregates of amyloid-beta peptides known as Amyloid Plaques (Hardy and Higgins, 1992) and intracellular aggregates of microtubular proteins called Neurofibrillary Tangles (Spillantini and Goedert, 1998) in the brain. This aggregation of amyloid-beta and microtubular associated protein Tau by formaldehyde in relation to Alzheimer's Disease pathology has been independently verified *in vitro* by three groups led by Dr. He (Lu et al., 2013; Nie et al., 2007a & 2007b), Dr. Peter Yu at the University of Saskatchewan (Chen et al., 2006 & 2007) and Dr. Xintian Hu also at the Chinese Academy of Sciences (Rizak et al., 2014). While it is important to note that formaldehyde induced aggregation tends to be a ubiquitous effect of formaldehyde at high concentrations on proteins, it is the unique toxicity of formaldehyde molecules to attack proteins that appears to have direct relevance to the pathology of Alzheimer's disease and to a number of other potential disorders (Henzi, 1984; Yu and Deng, 1998; Gubisne-Haberle et al., 2004).

Dr. Woodrow Monte, professor emeritus at the University of Arizona State, has postulated a number of roles for formaldehyde in the development of many disorders including Alzheimer's disease, Multiple Schlerosis, Lupus and Atheroschlerosis (Monte, 2011); all of which have been understudied in relation to formaldehyde. Dr. Monte recounts in *While Science Sleeps: A Sweetener Kills* a troubling story that might account for the lack of investigation by the main-stream scientific community into formaldehyde with respect to many of these diseases. Briefly, Dr. Monte describes a personal experience in reviewing data for the Food and Drug Administration (U.S.A.) into the safety of the sugar substitute aspartame, in which he refers to evidence warning against its use and a lobbying effort to ascertain approval. The concern regarding the use of aspartame revolved around a metabolic side product, methanol, a single carbon precursor of formaldehyde, and a number of toxic effects (reviewed in Monte, 2011). Nonetheless, the subsequent approval of aspartame by the FDA, unintentionally or not, has contributed in some degree to a lower importance for research into the effects of aspartame and its metabolites, methanol and formaldehyde, with respect to disease.

In hindsight, the recent developments into the effects of formaldehyde suggest that the decision regarding aspartame may have been premature. For example, Dr. Monte outlined a number of trends regarding the increasing prevalence of a number of diseases, referred to as "diseases of civilization," that follow the increased consumption of aspartame, such as in Alzheimer's disease. Dr. Monte also presented a number of intriguing aspects that are consistent with recent observations that suggest formaldehyde may be the detrimental end product of aspartame consumption linked to disease. These aspects include, but are not limited to, the differences in how differing animal model systems metabolize methanol in comparison to humans and how the U-shaped alcohol consumption curve relates to disease prevention (Monte, 2011). However, few of these aspects have received full inquiry by the scientific community in the past, and the full extent of the role of formaldehyde in the development of diseases like Alzheimer's disease remains inconclusive. A better and more thorough evaluation and assessment into the epidemiology of aspartame and its metabolism in differing systems will provide a better understanding. The fact that 30 years after the original approval of aspartame there are considerable lines of evidence that question the down-stream effects of aspartame or methanol consumption suggests that there is a need to re-evaluate dietary consumption with respect to formaldehyde toxicity and disease progression.

Recent investigations have identified multiple links that point to formaldehyde as being a central factor in the development of Alzheimer's disease. These include, in no particular order, the aforementioned aggregation of Alzheimer's related proteins (Chen et al., 2006 & 2007; Lu et al., 2013; Nie et al., 2007a & 2007b; Rizak et al., 2014), tau protein hyperphosphorylation (a marker of the AD pathological cascade) (Lu et al., 2011 & 2013; Yang et al., 2014b), up-regulation of genes that are involved in the disease's progression (Im et al., 2006), the inhibition of long-term potentiation (a cellular process of neurons that results in the strengthening of synapses which is widely considered a molecular foundation for memory formation)(Tong et al., 2013) and deficits in spatial memory and cognitive abilities (Malak et al., 2003; Lu et al., 2008; Tong et al., 2013). There is also further potential that molecular processes associated with Alzheimer's disease, such as the binding of amyloid peptides to mitochondrial Alcohol Dehydrogenase (Borger et al., 2013), contribute to formaldehyde accumulation in the brain. This suggests that there might be a feed-forward mechanism that augments disease progression. Consideration of these findings in conjunction with the finding that concentrations of formaldehyde increase with aging (Tong et al., 2013) highlight the potential importance of formaldehyde to Alzheimer's disease progression.

However, a better understanding of the implications of formaldehyde in Alzheimer's disease requires a full grounding in the scientific advances made with respect to the disease itself.

Alzheimer's disease was first described in 1906 by Alois Alzheimer (1864-1918) after a histological examination of a female patient that died following a decline in her mental capabilities. This investigation provided the first evidence of neuritic plaques, neuronal tangles and neuronal loss being associated with cognitive and memory impairments (reviewed in Martin, 2002). It was over 80 years later that these pathological markers were identified to be Amyloid Plaques, containing amyloid-beta peptides (Hardy and Higgins, 1992) and Neurofibrillary Tangles, containing hyperphosphorylated tau protein (Spillantini and Goedert, 1998). Nonetheless, the neuronal atrophy associated with these protein aggregates in the aging brain has become the main pathological marker of Alzheimer's disease as a neurodegenerative dementia. Typically, positive diagnosis of the disease has required post-mortem investigation of a patient's brain, but modern advancements in brain imaging has aided in identifying brain atrophy in Alzheimer's patients prior to death (Blennow et al., 2006; Ryan et al., 2013), giving hope to successful treatment of the disease. Unfortunately, the direct etiology of Alzheimer's disease, the neurodegeneration and the protein aggregation has remained elusive to date. This has limited development of more effective treatments that control the progression of the disease.

In addition to the aspect of aging, as first mentioned, in relation to Alzheimer's disease progression, genetic mutations have been identified as important factors to the development of the disease (reviewed in Bertram and Tanzi, 2012). The most notable mutations include a variation of mutations to a gene that codes for the Amyloid Precursor Protein. The Amyloid Precursor Protein is an extracellular membrane bound protein that undergoes an intricate cleavage process that results in the production of a number of smaller peptides, where mutations to the gene increase susceptibility for cleavage of the Amyloid Precursor Protein into a specific peptide, the amyloid-beta peptide (commonly between 40-42 amino acids), which tends to cluster and aggregate into extracellular plaques (reviewed in Bertam and Tanzi, 2012 & in Zhang and Song, 2013). The lineage of this mutation has been mapped to show inheritance of the disease within families. There is also another set of mutations in two other genes, Presenilin 1 & 2, that are inherited within families and contribute to Alzheimer's disease development. Interestingly, the Presenilin 1 and 2 proteins are associated within a protein complex that cleaves the Amyloid Precursor Protein and the mutations in Presenilin 1 or 2 are located within or near the cleavage sites, ultimately contributing to the presence of amyloid-beta in Alzheimer's disease. These mutations all contribute to inherited forms of Alzheimer's disease or familial Alzheimer's disease (Bertam and Tanzi, 2012). However, these inherited types of Alzheimer's disease only make up approximately 1 % of all diagnosed Alzheimer cases (Thies and Bleiler, 2013).

The predominant form of Alzheimer's disease is sporadic without a definitive causation. There are a number of other genes that are considered risk factors for the disease, but inheritance of any of these genetic risk factors do not correlate directly to disease onset (Bertam and Tanzi, 2012). This indicates there is a complex process that involves some of the gene mutations associated with sporadic Alzheimer's disease but is not limited to their presence and that additional factors are involved in the progression of the disease. One of the more interesting and important distinctions between familial and sporadic Alzheimer's disease is their age of onset. Familial Alzheimer's disease is also referred to as Early Onset Alzheimer's disease and appears clinically at younger ages (~65 years) than sporadic

Alzheimer's disease. Likewise, sporadic Alzheimer's disease is thus referred to as Late Onset Alzheimer's disease (Thies and Bleiler, 2013). It has not been fully elucidated as to how age contributes to the progression of the disease, but there is a new and strong impetus to consider formaldehyde to be a creditable factor.

The long standing model of Alzheimer's disease progression and pathology is referred to as the beta-amyloid cascade hypothesis (reviewed in Reitz, 2012). The beta-amyloid cascade hypothesis originated with the understanding that mutations to the Amyloid Precursor Protein gene lead to increased amyloid-beta peptide production in Early Onset Alzheimer's disease. From this, the increased content of amyloid-beta in the brain was described to both aggregate extracellularly into plaques and then to induce intracellular phosphorylation of the microtubular associated protein Tau. The hyperphosphorylation of Tau has been found to then destabilize the microtubular structure in neurons causing microtubules to disintegrate and, ultimately, to aggregate into neurofibrillary tangles in the cells. Tau hyperphosphorylation and the formation of neurofibrillary tangles are then considered initiators of cell death and the neuronal atrophy witnessed in Alzheimer's disease (Spillantini and Goedert, 1998; Obulesu et al., 2011). This cascade promotes the notion that amyloid plaque formation precedes neurofibrillary tangles formation in Alzheimer's disease pathological development (Reitz, 2012). More recently, a few studies have suggested that tau phosphorylation states may even modulate Amyloid Precursor Protein cleavage (Luan et al., 2013) or that the the two hallmarks (amyloid plaques and neurofibrillary tangles) may develop independently (Perani, 2014). Thereby, raising new questions into the direct etiology of Alzheimer's disease, especially as it pertains to sporadic forms. As such, new hypotheses, such as the amyloid-mitochondrial cascade, where beta-amyloid peptides interact with mitochondria causing irreversible damage, dysfunction and oxidative damage have been proposed (Morley and Farr, 2014). In any regard, these cascades are postulated to lead to neuronal dysfunction, cell death and the atrophy witnessed in Alzheimer's disease brains. It is less clear, however, how aging corresponds to the development of pathological hallmarks in sporadic or late onset Alzheimer's disease in the absence of the genetic instigators (mutations to the amyloid precursor protein or Presenilin 1 & 2) in familial AD that highlight the beginning of the cascade hypothesis.

Focus on genetic factors involved in the development of late onset Alzheimer's disease identified a number of mutations that convey higher risk. The best known and most important of these factors are mutations to the Apolipoprotein E (APOE) gene. Mutations to this gene correspond to two loci within the protein, coding for either cysteine or arginine at one or both loci, respectively (Betram and Tanzi, 2012). Ancestrally in lower order animals and common laboratory animal models, the Apolipoprotein contains arginines at the two loci (McIntosh et al., 2012). This corresponds to what is known as the APOE4 allele in humans, which codes for APOE proteins with arginines at residues 112 and 158. Humans also carry APOE alleles with cysteines at one or both of these loci. The APOE3 allele has a cysteine at residue 112 and an arginine at 158 and the APOE2 allele has a cysteine at both residue 112 and 158. The APOE3 allele is considered a disease neutral or natural state in humans with the risk variance of Alzheimer's disease increasing from APOE2 to APOE3 to APOE4 where APOE4 has a 400% higher relative risk for Alzheimer's disease development than the other alleles or corresponding protein isoforms (Bertram and Tanzi, 2012). In light of this pattern of vulnerability, there is a possibility, although speculative at this time, that these arginine to cysteine mutations confer a level of protection against cognitive decline that meets a new evolutionary challenge in the aging brain of humans. Conventionally, these mutations have been postulated to have arose in response to changes in dietary challenges between ancestral animals and humans because the predominant role of APOE is in cholesterol metabolism (McIntosh et al., 2012). However in Alzheimer's disease, APOE has been implicated in the clearance of amyloid-beta peptides from the extracellular space, where the APOE4 isoform tends to lead to more amyloid-beta accumulation in plaques (Castellano et al., 2011). Moreover, APOE has been found to be present in amyloid plaques examined post-mortem (Namba et al., 1991), suggesting that it has a direct, albeit inconclusive role, either positive or negative, in the formation of amyloid plaques in Alzheimer's disease.

The implications of these arginine residues in the APOE4 isoform and its reduced ability to clear amyloid-beta have been recently considered to have direct relevance to the findings of increased formaldehyde in the aging brain (Rizak et al., 2014). Formaldehyde as a small planar polar single carbon molecule has a unique reactivity that interacts with multiple agents. With amines, formaldehyde forms nitrogen-carbon covalent bonds that are often used in Mannich reactions during specific organic chemical synthesis. In protein biochemistry, formaldehyde attacks amines, in particular large amino acids with extended amino groups protruding from the tertiary structure of proteins such as lysine, tryptophan and arginine, forming adducts that might impair protein function (Toews et al., 2008; Edissi et al., 2013). Interestingly, protein carbonylation has been considered a biomarker of Alzheimer's disease, but has not been specifically associated to formaldehyde adducts (Greilberger et al., 2010). Nonetheless, such adducts are also free to interact with neighbouring or adjacent peptides to form methylene bridge cross-links (Toews et al., 2008). The arginine residues in APOE4 provide, in principle, differing sites of reactivity towards formaldehyde which coincides with the genetic vulnerability of Alzheimer's disease (Figure 1). Stable cross-links have been spontaneously formed between amyloid-beta peptides, APOE proteins and mixtures of the two in *in vitro* laboratory studies containing formaldehyde, leading to the formation of large aggregates of the proteins (Rizak et al., 2014).

The aggregation pattern of amyloid-beta with APOE isoforms *in vitro* reflected the trend of genetic vulnerability of APOE alleles in Alzheimer's disease. APOE4 formed more aggregates with amyloid-beta peptides in high concentrations of formaldehyde (10 mM) than with APOE3 or APOE2, respectively (Rizak et al., 2014). While there was a significant differential aggregation pattern observed, the studies were initial investigations utilizing aggregation formations visualized under denaturing conditions with western blots and it is not known whether this pattern holds in living systems. Further work is needed to evaluate this genetic relationship with formaldehyde using more sensitive apparatuses *in vitro* and, specifically, in animal systems. It should also be noted that the concentrations used in these aggregation assays were 10-100 fold higher than the formaldehyde concentrations reported in urine, cerebral spinal fluid and brain of Alzheimer's patients (Tong et al., 2013). However, amyloid-beta aggregation has also been witnessed *in vitro* at lower concentrations of formaldehyde (Chen et al., 2007; Rizak et al., 2014). In any regard, it has been difficult to assess the actual concentration of formaldehyde within the micro-environment that plaques are formed *in situ,* as well as the continued effect of having formaldehyde in the system over extended periods, such as those experienced during aging.

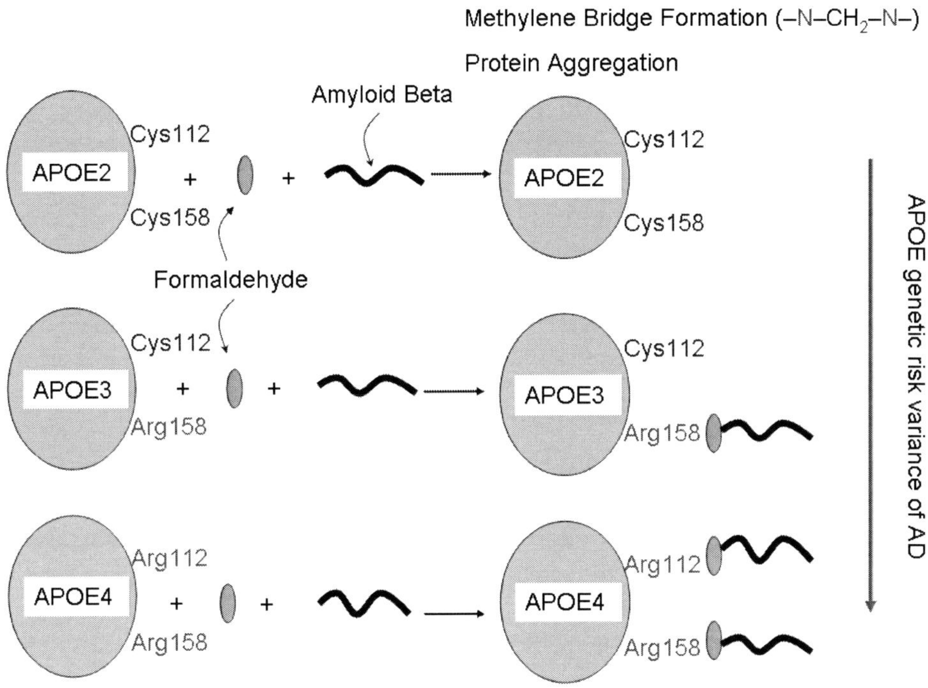

Figure 1. Schematic representation of the potential Formaldehyde interactions with APOE isoform mutations in the aggregation of Amyloid Beta. Reproduced from (Rizak, 2014).

Two modern studies performed by Dr. Xintian Hu's group in collaboration with Dr. Rongqiao He at the Chinese Academy of Sciences re-assessed the chronic effects of formaldehyde utilizing methanol feeding in mice and rhesus monkeys on Alzheimer's disease pathological markers and highlighted a distinction between laboratory animals assessing one-carbon toxicity (Yang et al., 2014a & 2014b). This distinction between laboratory animals was also stressed by Dr. Monte of Arizona State University in his evaluation of data on aspartame and methanol for the FDA (Monte, 2011). Interestingly, these two studies found that chronic methanol feeding induced Amyloid Plaque formation in monkeys, but not in mice (Yang et al., 2014a & 2014b). Methanol feeding was necessary over formaldehyde feeding because the animals would not consume formaldehyde *ad libitium* and abstained from consumption in these studies. While methanol is metabolized to formaldehyde and formic acid in both species, it is noteworthy that different animals utilize different mechanisms to metabolize methanol. Rodents, for example, are more adept at a detoxifying one-carbon moieties than primates (Johlin et al., 1987; Sweeting et al., 2010). Rodents have been postulated to metabolize methanol to formaldehyde with catalase in the peroxisomes limiting the damage from the toxic end-products, while primates have been suggested to predominantly metabolize methanol in competition with ethanol in the cytosol or mitochondria via Alcohol Dehydrogenase III (Monte, 2011). This phenomenon has been related to the U-shape consumption curve of ethanol, where no ethanol consumption leads to methanol metabolism and formaldehyde production and low but regular ethanol consumption prevents methanol metabolism by out competing for Alcohol Dehydrogenase III activity, limiting the toxic effects of formaldehyde and accounting for lower incidences of disease, such as Alzheimer's disease, in moderate alcohol drinkers (Monte, 2011). However, and like many other one-carbon toxicity theories, these avenues have not received the full attention of a scientific inquiry in the past, predominantly because earlier studies in rodents and other animals focused on the methanol metabolic end product, formate, which is quickly removed and often associated with acute toxicity (Hanston, 2006). Many studies did not return results that warranted extreme caution with regards to methanol or formaldehyde. Nonetheless, the two aforementioned studies in mice and monkeys, as well as *in vitro* comparisons of methanol metabolites which found that formaldehyde is more toxic in relation to Alzheimer's disease pathological markers than methanol or formate (Yang et al., 2014b), have brought attention to the fact that the distinction between laboratory animal choice and metabolite focus have implications that run beyond earlier studies and suggest that important information regarding health and safety may have been missed in the past as well as limited therapeutic development.

Although the monkeys and mice displayed differences in amyloid plaque development in response to methanol toxicity, both species displayed intracellular tau-phosphorylation and memory impairments. While it is not clear as to what distinction separates the animals other than differing metabolic processes to detoxify methanol, the toxic events in mice show the dangers of methanol or formaldehyde toward intracellular processes leading to cell loss in the brain have implications for Alzheimer's disease pathogenesis (Yang et al., 2014b). Unfortunately the mechanisms underlying these changes have not been well described to date. Nonetheless, both methanol fed models have unique qualities that can help elucidate these mechanisms. For example, the mice displayed a methanol toxicity that appears more related to down-stream intracellular tau-phosphorylation dependent pathways. It is possible that methanol metabolism in the peroxisome limits spread of formaldehyde and prevents it from interacting with the Amyloid Precursor Protein or amyloid-beta in the mice. The monkeys, on the other hand, displayed a methanol toxicity that presented with symptoms related to the three major pathological markers of Alzheimer's disease: Amyloid Plaque formation, tau hyperphosphorylation and memory impairments. This is, to the author's knowledge, the first evidence outlining a mechanism that induces all pathological markers of Alzheimer's disease from a single origin, i.e., one carbon toxicity.

More recently, Dr. Hu's group has re-investigated the effects of methanol and formaldehyde in the brain of rhesus macaques under more controlled conditions than through the *ad libitum* feeding of methanol. These recent studies utilized direct intracerebroventricular administration of either methanol or formaldehyde in adolescent monkeys over a 4 month period. Administration of formaldehyde directly to the lateral cerebral-ventricle in monkeys at concentrations comparable to concentrations found in Alzheimer's disease and aging populations (Tong et al., 2013) have found formation of extracellular aggregates of amyloid-beta (note: this data is preliminary and has not been peer reviewed or published to date). Conversely, direct administration of methanol to the lateral cerebral-ventricle displayed intracellular amyloid aggregates. The different locations of aggregation suggests that methanol is processed intracellularly, which then led to localized aggregation, where extracellular formaldehyde directly contributed to the aggregation of amyloid protein outside of the cells. These findings are consistent with intracellular methanol metabolism and indicate that extracellular formaldehyde contributes to extracellular plaque formation. This is important because it suggests that formaldehyde may not need be produced within the brain to cause Alzheimer's disease as formaldehyde is capable of passing through the blood brain barrier (Shcherbakova et al., 1986).

However, these studies were not able to assess the aggregation pattern of APOE isoforms and amyloid beta in the presence of formaldehyde *in situ* because natural model animals have a genetic make-up that parallels APOE4. While no monkey lines exist at present that express the varying alleles, transgenic mice lines are commercially available that carry one of the three different alleles. Original assessment of utilizing these transgenic mice lines with methanol feeding to investigate the differential aggregation of APOE isofoms with amyloid-beta was not judged to be conducive because the rodent system is less vulnerable to one carbon toxicity and the metabolic break-down of methanol does not produce notable amyloid plaque formation in mice (Yang et al., 2014b). However, the recent, albeit non-published, results obtained in monkeys by Dr. Hu's group with direct intracerebroventricular administration of formaldehyde suggests that direct administration of formaldehyde to APOE transgenic mice may provide a suitable background for assessing this differential aggregation of APOE isoforms and amyloid beta as long as expression levels of each isoform and amyloid-beta can be controlled or matched amongst groups. Establishing the trend of APOE genetic vulnerability with respect to formaldehyde induced amyloid aggregation in an animal system is important because it will provide a direct link between the most common genetic factor in sporadic Alzheimer's disease and the most common pathological factor – aging, as formaldehyde is an age-related marker (Wang et al., 2010; Tong et al., 2011 & 2013), in a complex biological environment that models the progression of the disease in the aging brain.

There is another link between formaldehyde and Alzheimer's disease that is concerned with the one-carbon cycle and has direct and important relevance. The one-carbon cycle sequesters one-carbon moieties within the brain, body and blood and plays a role in a number of molecular systems, such as in epigenetic regulation (reviewed in Fuso, 2013). Within the cycle, one-carbon moieties are transferred from tetrahydrofolate, folate and vitamin B12 to methionine and homocysteine, where S-adenosyl methionine is important methyl donor that plays a number of roles in cellular regulation. After methyl transfer, S-adenosyl methionine becomes S-adenosyl homocysteine (Figure 2). However, this one-carbon cycle appears to be compromised in Alzheimer's disease and the cycle shifts its balance in favor of a condition known as hyperhomocysteremia (Koseoglu and Karaman, 2007; Linnebank et al., 2010; Arlt et al., 2012). There are also a number of genes and proteins that are associated with the cycle that have been considered factors or markers of Alzheimer's disease, such as the 5,10-methylene tetrahydrofolate reductase gene (Postiglione et al., 2001), methionine synthase gene (Beyer et al., 2004) and the methionine adenosyl-transferase enzyme (Gomes Trolin et al., 1995). In any regard, free formaldehyde is believed to be spontaneously generated or captured from this cycle (Tyihak et al., 1998) and the fact that homocysteine (a non-peptidyl amino acid which lacks the side-chain carbon of methionine) levels are increased, S-adenosyl methionine levels are decreased and the S-adenosyl methionine to S-adenosyl homocysteine ratios are decreased in Alzheimer's disease (Linnebank et al., 2010), suggests that this cycle may contribute the excess formaldehyde witnessed in Alzheimer's patients. While hyperhomocysteremia has been well studied and correlated to the CSF of Alzheimer's patients and to biomarkers of Alzheimer's disease such as Tau phosphorylation (Arlt et al., 2012), no correlation study has ever been performed relating homocysteine levels with formaldehyde in Alzheimer's disease. Nonetheless, formaldehyde has been found to correlate with the same Tau phosphorylation biomarkers in animals (Yang et al., 2014a & 2014b). These findings of formaldehyde, hyperhomocysteremia and the compromised one-carbon cycle in Alzheimer's disease suggest that they are likely interrelated and have a presumably integrated role in the progression of the disease.

Moreover, the one-carbon cycle is a major source of the S-adenosyl homocysteine required for the production of glutathione, the major brain anti-oxidant. It is well accepted that glutathione levels decrease with aging (Kasapoglu and Ozben, 2001). Although it has not be fully evaluated, it is plausible that increasing formaldehyde levels with age reciprocate the decrease in glutathione levels in the aging brain as formaldehyde is known to interact with the thiol-group of glutathione and because it has been established that glutathione levels decrease in the presence of formaldehyde (Harris et al., 2004).

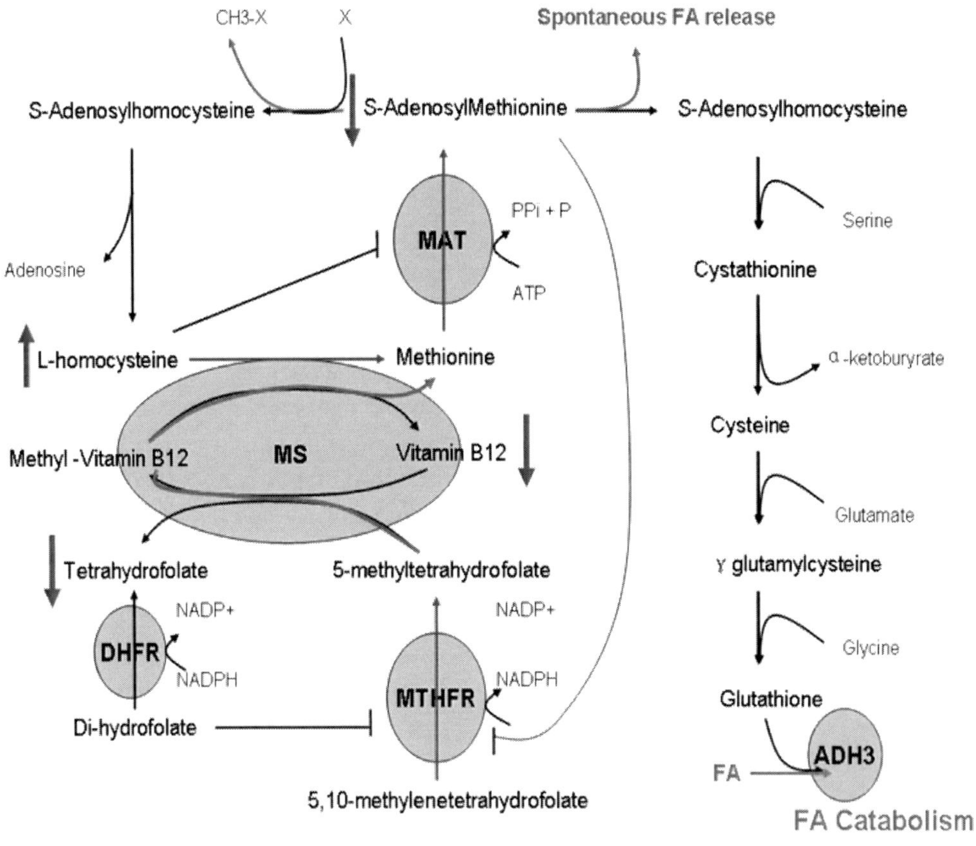

Figure 2. Schematic representation of the one carbon cycle as it pertains to Alzheimer's Disease (left side) and the glutathione synthesis pathway (right side). Green Arrows indicate metabolites that have altered levels in AD. Red Arrows indicate enzymes that have potential functional alterations in AD. Blue Arrows indicate the path of one carbon moieties. Arrows with blunt ends (⊣) represent inhibition. Enzyme Abbreviations: MS methionine synthase; DTHR dihydrofolate reductase; MTHFR 5,10, methylene tetrahydrofolate reductase; MAT ATP dependent: 1-methionine S-adensyl transferase; ADH3 glutathione dependent - alcohol dehydrogenase 3. Reproduced from (Rizak, 2014).

Another potential mechanism related to age associated increases in formaldehyde is the global demethylation of DNA that occurs during aging (Liu et al., 2009; Wu and Zhang, 2010). Epigenetic regulation via cytosine methylation, where S-adenosyl Methionine supplies the methyl group, has important roles in regulating a wide-array of gene functions and has been implicated to have a role in the aging process (Berdasco and Esteller, 2012). However, there have been no direct correlations studies to connect DNA demethylation to the increased formaldehyde witnessed by Dr. He's group in aging populations. Nonetheless, an intricate balance of one-carbon moieties appears to be necessary for maintaining a number of functions, such that any such global change affecting a major store of methyl groups, such as methylated DNA, may contribute directly to free one-carbon moieties in the blood or cerebral spinal fluid.

The full role of a healthy and balanced one-carbon system has not received a fair evaluation by the scientific community and may be involved in a number of pathogenic states beyond Alzheimer's disease. For example, hyperhomocysteremia is correlated to both vascular disease (Nilsson et al. 2008) and vascular dementia (Koseoglu and Karaman, 2007). Interestingly, Dr. Yu's group at the University of Saskatchewan has also postulated a link between formaldehyde in vascular disease and vascular dementia. However, Dr. Yu's group found the increases of formaldehyde in an atherosclerotic mouse model to be caused by the methylamine catalysis of semicarbazide-sensitive amine oxidase (SSAO) (Gubisne-Haberle et al., 2004; Jiang et al., 2008), rather than related to homocysteine levels. Interestingly, SSAO has been found to be over-expressed in Alzheimer's disease (Unzeta et al., 2007) and to co-localize with amyloid plaques (Jiang et al., 2008), suggesting that formaldehyde from a number of sources may contribute to disease progression. Although this postulation did not focus directly on a relationship to the one-carbon cycle, formaldehyde toxicity, especially its ability to form protein cross-links, has direct relevance to a number of diseases, including atherosclerosis, vascular dementia and Alzheimer's disease.

The multiple endogenous sources of formaldehyde or one-carbon moieties found within the body provide a strong impetus for the medical community to research further, not only because there are direct connections between these sources and Alzheimer's disease but because formaldehyde as an exogenous toxin provides an equally strong risk towards disease. This is the threat that Dr. Monte brought to light in outlining the potential risks of aspartame, as well as a number of other methanol sources common to diet, such as canned fruits and vegetables (Monte, 2011). While the full correlation between these examples and Alzheimer's disease and other diseases hasn't been elucidated, the chemical nature of formaldehyde with respect to disease pathology warrants careful re-evaluation of these concerns such that better understanding can be heeded in the future. For example, cigarette and tobacco smoking have been considered risk factors for Alzheimer's disease (Chang et al., 2012) and both aspirants contain levels of methanol and formaldehyde (Baker, 2006). However, no investigation to date has been directed at or been able to directly correlate formaldehyde and its protein cross-linking ability to the progression of the disease. It is clear from the work of Dr. Hu's group that chronic methanol toxicity has the potential to cause Alzheimer's disease pathological markers in primates (Yang et al., 2014a). However, the fact that this toxicity only presents with partial pathological markers in rodents suggest that previous investigations in rodents may have not been able to fully engage the entire spectrum of Alzheimer's disease development in those investigations.

This confound may have limited a number investigations designed to identify a number of drugs for the treatment of Alzheimer's disease. Unfortunately, most drugs developed in rodent systems often fail in clinical human trials (Arrowsmith, 2011; Panza et al., 2011). It is possible that the differential one-carbon metabolism between rodents and primates plays a negative role in these investigations. It is notable that rodents do not develop Amyloid Plaques unless the animals have been genetically modified to over-express the Amyloid Precursor Protein or related mutants (McCampbell et al., 2011; Lu et al., 2013). It is possible that treatments that reduce Amyloid Plaque formation in these type of models limit the effect of amyloid aggregation by modulating protein amounts, rather than treating a fundamental property that rodents naturally protect against by metabolizing one-carbon moieties within the peroxisome. In humans, these types of treatments may only provide benefits in cases of early onset or inherited Alzheimer's disease where altered protein processing contributes to plaque formation, while in sporadic Alzheimer's disease the specific impairment that leads to aggregation remains elusive and goes untreated.

Although it has not been established that formaldehyde is such a factor, there are a number of potentials, as reviewed here, that warrant its consideration. To establish this, however, will require detection of formaldehyde specifically within plaques and tangles associated with Alzheimer's disease. This is a great challenge, not only because most post-mortem tissue is fixed in para-formaldehyde before investigation but because detection of formaldehyde and formaldehyde adducts, as a single carbon addition to proteins, is difficult. Moreover, existing antibodies raised against formaldehyde adducts have been found to cross react with acetylaldehyde adducts (Pietrzak et al., 1995), limiting their effectiveness. Experimentally, animals administered (C-14) labeled formaldehyde may provide fruitful knowledge that can demonstrate radioactive signatures originating specifically from within plaques or tangles. Although such experiments would be able to demonstrate a direct role of formaldehyde in aggregation rather than any other potential down-stream effect of formaldehyde, the technical challenges of working with radioisotopes and animals limit these experiments to appropriate institutions. Nonetheless, the direct confirmation of formaldehyde cross-links in Amyloid Plaques and Neurofibrillary Tangles would, without a doubt, establish a better understanding of diseases like Alzheimer's, where aggregation is paramount to the pathological progression, as well as open new avenues for treatment.

Interestingly, there is a natural product and anti-oxidant commonly found in grapes and red wine, resveratrol, that is currently being investigated as an Alzheimer's disease agent (Karuppagounder et al., 2009; Porquet et al., 2014). Resveratrol, a polyphenol compound, gets its fame, in part, from what is known as the "French Paradox" in relation to the regular consumption of red wine in French culture and its potential health benefits (Tyihak et al., 2011). These benefits are synonymous with the U-shape consumption curve of alcohol that Dr. Monte highlighted in relation to methanol metabolism. Resveratrol garners most of its attention as an anti-oxidant through the up-regulation of genes like Sirituin-1 in the protection of cells (Porquet et al., 2014). However, resveratrol has also been implicated to directly act as formaldehyde scavenger and carrier molecule (Tyihak et al., 1998 & 2011). There is one caveat worth mentioning with regard to the recent investigations of resveratrol. Although studies have found reduce amyloid plaque pathology after resveratrol treatment (Karuppagounder et al., 2009), resveratrol has been found to bind amyloid beta directly (Ge et al., 2012). In an unpublished *in vitro* study in Xintian Hu's lab, it was found that resveratrol binds amyloid beta irreversibly in the presence, but not absence, of formaldehyde, which impairs anti-amyloid-beta antibody specificity, most likely by altering the binding epitope. This confound may have implications with respect to earlier studies that investigated the effect of resveratrol on amyloid-beta levels that warrants further consideration in future studies. At this point, it is not clear what role resveratrol plays, if any, in the treatment of

Alzheimer's disease, but the implications that formaldehyde is directly involved in amyloid and tau aggregation provides hope that resveratrol has a role to play in the treatment of Alzheimer's disease through the sequestering of free one-carbon moieties in the blood or cerebral spinal fluid.

While there has been no model established into how formaldehyde is involved, either as a sole cause or a contributing factor, to the development of Alzheimer's disease, there are multiple situations, outlined here, that point to formaldehyde being an important aspect, if not the central aspect, in the development of the disease. It seems that the most common concern of the scientific and public domains with respect to formaldehyde or methanol has been focused on exogenous exposure to the agent, as those found in histologists with impaired memory in relation to workplace exposure (Kilburn et al., 1987). Although this finding keenly relates to a major symptom of Alzheimer's disease and it is sincerely important that all areas of society be aware of the potential effects of exogenous formaldehyde and related sources, as Dr. Monte points out with the metabolism of aspartame, to the development of Alzheimer's disease, it is the natural correlation of the disease to formaldehyde that occurs endogenously which might warrant more considerable attention by the medical research community going forward. These points of progression include the natural increase of formaldehyde levels with aging, the aspects of protein aggregation and protein cross-linking caused by formaldehyde, the potential relationship of formaldehyde to APOE genetic vulnerability and the disturbances of the one-carbon cycle, such as hyperhomocysteremia, related to Alzheimer's disease.

It is because of all of these aspects that has posed the author, in collaboration with Dr. Xintian Hu, to ask the question "Is formaldehyde the missing *link* in Alzheimer's disease pathology?"(Rizak et al., 2014). This question is more than a play on the cross-linking abilities of formaldehyde, a phenomenon well-established in scientific experimentation, it is the unique characteristics of formaldehyde which unify a number of observations found in Alzheimer's disease that have otherwise seemed unrelated to date. Although the question has not been fully answered, it appears that it is the correct question to ask at this point. Alzheimer's disease is a poignant socioeconomic burden that is expected to affect 70 million people worldwide by 2030 with dire consequences for both families and economies across the globe (Thies and Bleiler, 2013). It has no definitive treatment or cure and the exact etiology of the disease remains unknown. The consideration of formaldehyde, for all the reasons listed above, has not only opened a debate on its implications in Alzheimer's disease progression and other related aggregation containing disorders, but has revitalized hope that future treatments for Alzheimer's disease based on mitigating metabolic processes related to formaldehyde can be attained that are separate, or in addition, to approaches focusing on genetic modulation.

ACKNOWLEDGMENTS

The author would like to thank Dr. Xintian Hu for discussions on the topic and for consenting to disseminate his labs most recent findings before publication elsewhere. Zhai RongWei was lead experimenter investigating methanol and formaldehyde through intracerebroventricular administration. Margaret Kellogg, a high school summer student, and He XiaPing provided the work identifying that resveratrol binding amyloid-beta in formaldehyde altered antibody specificity.

REFERENCES

Arlt S, Schwedhelm E, Kolsch H, Jahn H, Linnebank M, Smulders Y, Jessen F, Boger RH, Popp J (2012) Dimethylarginines, homocysteine metabolism, and cerebrospinal fluid markers for Alzheimer's disease. *J Alzheimers Dis* 31:751-758.

Arrowsmith J (2011) Trial watch: Phase II failures: 2008-2010. *Nature Reviews Drug Discovery* 10:328-329.

Baker RR (2006) The generation of formaldehyde in cigarettes--Overview and recent experiments. *Food Chem Toxicol* 44:1799-1822.

Berdasco M, Esteller M (2012) Hot topics in epigenetic mechanisms of aging: 2011. *Aging Cell* 11:181-186.

Bertram L, Tanzi RE (2012) The genetics of Alzheimer's disease. Progress in Molecular Biology and Translational Science 107:79-100.

Beyer K, Lao JI, Latorre P, Riutort N, Matute B, Fernandez-Figueras MT, Mate JL, Ariza A (2003) Methionine synthase polymorphism is a risk factor for Alzheimer disease. *Neuroreport* 14:1391-1394.

Blennow K, de Leon MJ, Zetterberg H (2006) Alzheimer's disease. *Lancet* 368:387-403.

Bogdanffy MS, Morgan PH, Starr TB, Morgan KT (1987) Binding of formaldehyde to human and rat nasal mucus and bovine serum albumin. *Toxicology Letters* 38:145-154.

Borger E, Aitken L, Du H, Zhang W, Gunn-Moore FJ, Yan SS (2013) Is amyloid binding alcohol dehydrogenase a drug target for treating Alzheimer's disease? *Current Alzheimer Research* 10:21-29.

Castellano JM, Kim J, Stewart FR, Jiang H, DeMattos RB, Patterson BW, Fagan AM, Morris JC, Mawuenyega KG, Cruchaga C, Goate AM, Bales KR, Paul SM, Bateman RJ, Holtzman DM (2011) Human apoE isoforms differentially regulate brain amyloid-beta peptide clearance. *Science Translational Medicine* 3 (89):89ra57.

Chang CC, Zhao Y, Lee CW, Ganguli M (2012) Smoking, death, and Alzheimer disease: a case of competing risks. *Alzheimer Dis Assoc Disord* 26:300-306.

Chen K, Kazachkov M, Yu PH (2007) Effect of aldehydes derived from oxidative deamination and oxidative stress on beta-amyloid aggregation; pathological implications to Alzheimer's disease. *Journal of Neural Transmission* 114:835-839.

Chen K, Maley J, Yu PH (2006) Potential inplications of endogenous aldehydes in beta-amyloid misfolding, oligomerization and fibrillogenesis. *J Neurochem* 99:1413-1424.

Edrissi B, Taghizadeh K, Dedon PC (2013) Quantitative analysis of histone modifications: formaldehyde is a source of pathological n(6)-formyllysine that is refractory to histone deacetylases. *PloS Genet* 9(2):e1003328.

Fuso A (2013) The 'golden age' of DNA methylation in neurodegenerative diseases. *Clin Chem Lab Med* 51(3):523-34.

Ge JF, Qiao JP, Qi CC, Wang CW, Zhou JN (2012) The binding of resveratrol to monomer and fibril amyloid beta. *Neurochem Int* 61:1192-1201.

Gomes Trolin C, Regland B, Oreland L (1995) Decreased methionine adenosyltransferase activity in erythrocytes of patients with dementia disorders. *Eur Neuropsychopharmacol* 5:107-114.

Greilberger J, Fuchs D, Leblhuber F, Greilberger M, Wintersteiger R, Tafeit E (2010) Carbonyl proteins as a clinical marker in Alzheimer's disease and its relation to tryptophan degradation and immune activation. *Clin Lab.* 56(9-10): 441-8.

Gubisne-Haberle D, Hill W, Kazachkov M, Richardson JS, Yu PH (2004) Protein cross-linkage induced by formaldehyde derived from semicarbazide-sensitive amine oxidase-mediated deamination of methylamine. *J Pharmacol Exp Ther* 310:1125-1132.

Hantson PE (2006) Acute methanol intoxication: physiopathology, prognosis and treatment. *Bull Mem Acad R Med Belg* 161:425-434.

Hardy JA, Higgins GA (1992) Alzheimer's disease: the amyloid cascade hypothesis. *Science* 256:184-185.

Harris C, Dixon M, Hansen JM (2004) Glutathione depletion modulates methanol, formaldehyde and formate toxicity in cultured rat conceptuses. *Cell Biol Toxicol* 20:133-145.

He R, Lu J, Miao J (2010) Formaldehyde stress. *Science China Life Sciences* 53:1399-1404.

Henzi H (1984) Chronic methanol poisoning with the clinical and pathologic-anatomical features of multiple sclerosis. *Med Hypotheses* 13:63-75.

Im H, Oh E, Mun J, Khim JY, Lee E, Kang HS, Kim E, Kim H, Won NH, Kim YH, Jung WW, Sul D (2006) Evaluation of toxicological monitoring markers using proteomic analysis in rats exposed to formaldehyde. *Journal of Proteome Research* 5:1354-1366.

Jiang ZJ, Richardson JS, Yu PH (2008) The contribution of cerebral vascular semicarbazide-sensitive amine oxidase to cerebral amyloid angiopathy in Alzheimer's disease. *Neuropathol Appl Neurobiol* 34:194-204.

Johlin FC, Fortman CS, Nghiem DD, Tephly TR (1987) Studies on the role of folic acid and folate-dependent enzymes in human methanol poisoning. *Mol Pharmacol* 31:557-561.

Karuppagounder SS, Pinto JT, Xu H, Chen LH, Beal MF, Gibson GE (2009) Dietary supplementation with resveratrol reduces plaque pathology in a transgenic model of Alzheimer's disease. *Neurochem Int* 54:111-118.

Kasapoglu M, Ozben T (2001) Alterations of antioxidant enzymes and oxidative stress markers in aging. *Exp Gerontol* 36:209-220.

Kilburn KH, Warshaw R, Thornton JC (1987) Formaldehyde Impairs Memory, Equilibrium, and Dexterity in Histology Technicians - Effects Which Persist for Days after Exposure. *Arch Environ Health* 42:117-120.

Koseoglu E, Karaman Y (2007) Relations between homocysteine, folate and vitamin B12 in vascular dementia and in Alzheimer disease. *Clin Biochem* 40:859-863.

Linnebank M, Popp J, Smulders Y, Smith D, Semmler A, Farkas M, Kulic L, Cvetanovska G, Blom H, Stoffel-Wagner B, Kolsch H, Weller M, Jessen F (2010) S-adenosylmethionine is decreased in the cerebrospinal fluid of patients with Alzheimer's disease. *Neurodegener Dis* 7:373-378.

Liu L, van Groen T, Kadish I, Tollefsbol TO (2009) DNA methylation impacts on learning and memory in aging. *Neurobiol Aging* 30:549-560.

Lu Z, Li CM, Qiao Y, Yan Y, Yang X (2008) Effect of inhaled formaldehyde on learning and memory of mice. *Indoor Air* 18:77-83.

Lu J, Miao J, Pan R, He R (2011) Formaldehyde-mediated hyperphosphorylation disturbs the interaction between Tau protein and DNA. *Progress in Biochemistry and Biophysics* 38:1113-1120.

Lu J, Miao J, Su T, Liu Y, He R (2013) Formaldehyde induces hyperphosphorylation and polymerization of Tau protein both in vitro and in vivo. *Biochim Biophys Acta* 1830:4102-4116.

Luan K, Rosales JL, Lee KY (2013) Viewpoint: Crosstalks between neurofibrillary tangles and amyloid plaque formation. *Ageing research reviews* 12:174-181.

Martin JB (2002) The integration of neurology, psychiatry, and neuroscience in the 21st centrury. *American Journal of Psychiatry* 159:695-704.

Malek FA, Moritz KU, Fanghanel J (2003) A study on the effect of inhalative formaldehyde exposure on water labyrinth test performance in rats. *Ann Anat* 185:277-285.

McCampbell A, Wessner K, Marlatt MW, Wolffe C, Toolan D, Podtelezhnikov A, Yeh S, Zhang R, Szczerba P, Tanis KQ, Majercak J, Ray WJ, Savage MJ (2011) Induction of Alzheimer's like changes in brain of mice expressing mutant APP fed excess methionine. *J Neurochem* 116:82-92.

McIntosh AM, Bennett C, Dickson D, Anestis SF, Watts DP, Webster TH, Fontenot MB, Bradley BJ (2012) The apoliprotein E (APOE) gene appears functionally monoporphic in Chimpanzees (Pan troglodytes). *PloS one* 7(10): e47760.

Monte WC (2011) While Science Sleeps: A Sweetener Kills. San Francisco, USA: Amazon Create Space Publishing.

Morley JE, Farr SA (2014) The role of amyloid-beta in the regulation of memory. *Biochemical Pharmacology* 88:479-485.

Namba Y, Tomonaga M, Kawasaki H, Otomo E, Ikeda K (1991) Apolipoprotein E immunoreactivity in cerebral amyloid deposits and neurofibrillary tangles in Alzheimer's disease and kuru plaque amyloid in Creutzfeldt-Jakob disease. *Brain Res* 541:163-166.

Nie CL, Wang XS, Liu Y, Perrett S, He RQ (2007a) Amyloid-like aggregates of neuronal tau induced by formaldehyde promote apoptosis of neuronal cells. *BMC neuroscience* 8:9.

Nie CL, Wei Y, Chen X, Liu YY, Dui W, Liu Y, Davies MC, Tendler SJ, He RQ (2007b) Formaldehyde at low concentration induces protein tau into globular amyloid-like aggregates in vitro and in vivo. *PloS one* 2:e629.

Nilsson K, Gustafson L, Hultberg B (2008) Plasma homocysteine and vascular disease in elderly patients with mental illness. *Clin Chem Lab Med* 46:1556-1561.

Obulesu M, Venu R, Somashekhar R (2011) Tau mediated neurodegeneration: an insight into Alzheimer's disease pathology. *Neurochem Res* 36:1329-1335.

Orlando V (2000) Mapping chromosomal proteins in vivo by formaldehyde-crosslinked-chromatin immunoprecipitation. *Trends Biochem Sci.* 25(3): 99-104.

Panza F, Frisardi V, Imbimbo BP, Seripa D, Solfrizzi V, Pilotto A (2011) Monoclonal antibodies against beta-amyloid (A beta) for the treatment of Alzheimer's disease: the A beta target at a crossroads. *Expert Opin Biol Th* 11:679-686.

Perani D (2014) FDG-PET and amyloid PET imaging: the diverging paths. *Current Opinions in Neurology* 27:405-413.

Pietrzak ER, Shanley BC, Kroon PA (1995) Antibodies made against a formaldehyde-protein adduct cross react with an acetaldehyde-protein adduct. Implications for the origin of antibodies in human serum which recognize acetylaldehyde-protein adducts. *Alcohol Alcohol* 30:373-378.

Porquet D, Grinan-Ferre C, Ferrer I, Camins A, Sanfeliu C, Del Valle J, Pallas M (2014) Neuroprotective Role of Trans-Resveratrol in a Murine Model of Familial Alzheimer's Disease. *Journal of Alzheimer's disease* 42(4):1209-1220.

Postiglione A, Milan G, Ruocco A, Gallotta G, Guiotto G, Di Minno G (2001) Plasma folate, vitamin B(12), and total homocysteine and homozygosity for the C677T mutation of the 5,10-methylene tetrahydrofolate reductase gene in patients with Alzheimer's dementia. A case-control study. *Gerontology* 47:324-329.

Preston AR, Eichenbaum H (2013) Interplay of hippocampus and prefrontal cortex in memory. *Current Biology* 23:764-773.

Reitz C (2012) Alzheimer's disease and the amyloid cascade hypothesis: a critical review. *International Journal of Alzheimer's Disease* 2012: 369808.

Rizak JD (2014) The implications of formaldehyde in Alzheimer's Disease pathology. Doctoral Thesis, University of the Chinese Academy of Sciences.

Rizak JD, Ma Y, Hu X (2014) Is formaldehyde the missing link in AD pathology? The differentiation aggregation of amyloid-beta and APOE isoforms in vitro. *Current Alzheimer Research* 11:461-468.

Ryan NS, Keihaninejad S, Shakespeare TJ, Lehmann M, Crutch SJ, Malone IB, Thornton JS, Mancini L, Hyare H, Yousry T, Ridgway GR, Zhang H, Modat M, Alexander DC, Rossor MN, Ourselin S, Fox NC (2013) Magnetic resonance imaging evidence for presymptomatic change in thalamus and caudate in familial Alzheimer's disease. *Brain* 136:1399-1414.

Shcherbakova LN, Tel'pukhov VI, Trenin SO, Bashilov IA, Lapkina TI (1986) Permeability of the blood-brain barrier to intra-arterial formaldehyde. *Biull Eksp Biol Med* 102:573-575.

Spillantini MG, Goedert M (1998) Tau protein pathology in neurodegenerative diseases. *Trends Neurosci* 21:428-433.

Stuchlik A (2014) Dynamic learning and memory, synaptic plasticity and neurogenesis: an update. *Frontiers in Behaviour Neuroscience* 8:106.

Sweeting JN, Siu M, McCallum GP, Miller L, Wells PG (2010) Species differences in methanol and formic acid pharmacokinetics in mice, rabbits and primates. *Toxicology and Applied Pharmacology* 247:28-35.

Thies W, Bleiler L (2013) 2013 Alzheimer's disease facts and figures. *Alzheimer's & Dementia* 9:208-245.

Toews J, Rogalski JC, Clark TJ, Kast J (2008) Mass spectrometric identification of formaldehyde-induced peptide modifications under in vivo protein cross-linking conditions. *Analytica Chimica Acta* 618:168-183.

Tong ZQ, Zhang J, Luo W, Wang W, Li F, Li H, Luo H, Lu J, Zhou J, Wan Y, He R (2011) Urine formaldehyde level is inversely correlated to mini mental state examination scores in senile dementia. *Neurobiology of Aging* 32:31-41.

Tong ZQ, Han C, Luo W, Wang X, Li H, Luo H, Zhou J, Qi J, He R (2013) Accumulated hippocampal formaldehyde induces age-dependent memory decline. *Age* 35:583-596.

Tyihak E, Albert L, Nemeth ZI, Katay G, Kiraly-Veghely Z, Szende B (1998) Formaldehyde cycle and the natural formaldehyde generators and capturers. *Acta Biol Hung* 49:225-238.

Tyihak E, Kiraly-Veghely Z, Moricz AM (2011) Multiple beneficial effects of resveratrol and their chemical-biochemcial basis. *Natural Products Communications* 6:631-638.

Unzeta M, Sole M, Boada M, Hernandez M (2007) Semicarbazide-sensitive amine oxidase (SSAO) and its possible contribution to vascular damage in Alzheimer's disease. *J Neural Transm* 114:857-862.

Wang W, Hao Z, Zhang L, Li W, Li M, Zhang J, He R (2010) Research on urine formaldehyde concentration in Alzheimer's elderly and normal elderly. *Chin J Geriatr Heart Brain Vessel Dis* 12:721-722.

Wu SC, Zhang Y (2010) Active DNA demethylation: many roads lead to Rome. *Nat Rev Mol Cell Biol* 11:607-620.

Yang M, Miao J, Rizak JD, Zhai R, Wang Z, Huma T, Li T, Zheng N, Wu S, Zheng Y, Fan X, Yang J, Wang J, Yang S, Ma Y, Lü L, He R, Hu X (2014a) Alzheimer's Disease and Methanol Toxicity (Pt.2): Lessons from four rhesus macaques (macaca mulatta) chronically fed methanol. *Journal of Alzheimer's Disease* 41:1131-1147.

Yang M, Lu J, Miao JY, Rizak JD, Yang JZ, Zhai RW, Zhou J, Qu JG, Wang JH, Yang S, Ma YY, Hu XT, He RQ (2014b) Alzheimer's Disease and Methanol Toxicity (Part I): Chronic methanol feeding led to memory impairments and tau hyperphosphorylation in AD in mice. *Journal of Alzheimer's Disease* 41:1117-1129.

Yu PH, Deng YL (1998) Endogenous formaldehyde as a potential factor of vulnerability of atherosclerosis: involvement of semicarbazide-sensitive amine oxidase-mediated methylamine turnover. *Atherosclerosis* 140:357-363.

Zararsiz I, Kus I, Ogeturk M, Akpolat N, Kose E, Meydan S, Sarsilmaz M (2007) Melatonin prevents formaldehyde-induced neurotoxicity in prefrontal cortex of rats: an immunohistochemical and biochemical study. *Cell Biochem Funct* 25:413-418.

Zhang X, Song WL (2013) The role of APP and BACE1 trafficking in APP processing and amyloid-B generation. *Alzheimer's Research and Therapy* 5:46-48.

In: Formaldehyde
Editor: Amy Patton
ISBN: 978-1-63482-412-5

Chapter 5

LOW CONCENTRATION FORMALDEHYDE: POTENTIAL HEALTH EFFECTS AND TREATMENT TECHNOLOGY

Wenjun Liang, Jian Li and Hong He
Key Laboratory of Beijing on Regional Air Pollution Control,
Beijing University of Technology, Beijing, China

ABSTRACT

VOCs exist widely in both indoor & outdoor environment, and have been known as the most hazardous materials which affect public health and welfare owing to their toxicity potential, carcinogenicity and stability. VOCs are recognized as causative agents of the sick-building syndrome, in which formaldehyde (HCHO) is one of them. HCHO has been classified as a suspected carcinogen. HCHO is known to cause nasal tumors, irritation of the mucous membranes of the eyes and respiratory tract, and skin irritation. Because of the increasing concern on HCHO in the indoor environment, the abatement of HCHO is of significant practical interest at low temperature, especially at room temperature.

The conventional technique to remove VOCs is activated-carbon adsorption method, which needs to replace and dispose the adsorbent. The disposal of activated-carbon will cause the next environmental problem. The alternative methods are ozone oxidation method, catalytic combustion method, non-thermal plasma method, etc.. The ozone oxidation and the non-thermal plasma methods involve NOx emission and high initial cost. The catalytic combustion method requires a heat source, hence it is a highly energy-wasting process when VOC concentrations are low.

In this chapter, two kinds of HCHO decomposition technology were investigated, that is, photo-catalytic degradation and non-thermal plasma technology. Photo-catalytic degradation of VOCs on UV-illuminated titanium dioxide (TiO_2) is proposed as an alternative advanced oxidation process for the purification of water and air. Non-thermal plasma (NTP) has great industrial potential and has been applied to the decomposition of VOCs. At the same time, NTP technology has disadvantages such as low energy efficiency and undesirable byproducts such as NOx. So the combination of NTP with photocatalyst technology was used in our investigation.

Nano-structured TiO_2, Ag/TiO_2 and Ce/TiO_2 thin films were prepared by sol-gel method at room temperature and used in the photo-catalyst reactor for HCHO decomposition. The effects of doped $Ag/Ce/TiO_2$, relative humidity, oxygen concentration, initial HCHO concentration, UV light wave length, TiO_2 amount on HCHO decomposition were investigated. At the same time, destruction of formaldehyde by means of $NaNO_2$ ferro-electric packed bed dielectric barrier discharge plasma in a coaxial cylindrical reactor was carried out at atmospheric pressure and room temperature. The difference among four kinds of $NaNO_2$ ferro-electric reactors was compared in terms of specific energy density (SED), energy yield (EY), and HCHO decomposition. Three kinds of catalysts, that is, TiO_2, $BaTiO_3$, or TiO_2 & $BaTiO_3$ were tested for HCHO removal. Effect of catalyst carrier, catalysts, and catalyst amount on specific energy density (SED), energy yield (EY) and HCHO removal efficiency were investigated. In addition, the by-products were detected and the mechanism of HCHO degradation was performed.

INTRODUCTION

1. Potential Health Effects of Formaldehyde

Formaldehyde (HCHO) is an important chemical for the global economy, widely used in construction, wood processing, furniture, textiles, carpeting, and in the chemical industry. The 40% aqueous solution is termed formarin. Formaldehyde has a good reduction power that causes the degradation of cytoplasm and the coagulation of protein. Distilled formaldehyde solution is used for disinfection and antisepsis because of its bactericidal action. Although formaldehyde is a natural metabolic product of the human body, high-dose exposure increases the risk of acute poisoning, while prolonged exposure can lead to chronic toxicity and even cancer [1]. Formaldehyde levels between 0.12 and 0.6 mg/m^3 are detectable by human senses, between 0.6 and 1.2 mg/m^3 can cause eye irritation, and above 1.23 mg/m^3 can irritate the nose and throat [2].

Recently, many cases of poisoning, allergy, asthma, pulmonary damage, cancer and death were reported as a result of formaldehyde exposure from contaminated foods, drinking water, and polluted indoor air. It has been classified as a human carcinogen that causes nasopharyngeal cancer and probably leukemia. Formaldehyde is toxic to humans. In the case of a human being exposed to a low concentration of formaldehyde gas, the formaldehyde is pungent to the mucosa of the eye, nose, and respiratory tract and acts as a lachrymator causing sneezing and coughing.

It has been shown that the likelihood for the development of allergic asthma increases proportionately with indoor formaldehyde concentration, especially when levels exceed 0.12 mg/m^3 [3]. Long-term exposure to elevated levels of formaldehyde in the occupational setting has resulted in upper and lower airway irritation; eye irritation; and degenerative, inflammatory and hyperplastic changes of the nasal mucosa in humans [4-5]. It has been shown that formaldehyde exposure induces DNA and chromosomal damage in human peripheral blood cells [6]. Formaldehyde has been tested for carcinogenicity by inhalation, oral administration, topical application, and subcutaneous injections in rodents. It has been proposed that, based on the weight of evidence from in vivo studies, the likely mode of action for formaldehyde-induced nasal tumors in animals is relevant to humans at least qualitatively [7].

Based on comprehensive research and large-scale human studies conducted internationally, the International Agency for Research on Cancer (IARC) recently classified formaldehyde as a human carcinogen that causes nasopharyngeal cancer.

2. Conventional Technique of Removal of VOCs

The conventional technique to remove VOCs included such as carbon bed adsorption, absorption, condensation, and catalytic combustion.

Carbon bed adsorption, or adsorption by another material, is a process that does not convert the VOCs to other compounds as part of the process. Adsorption is an effective way of reducing the concentration of VOCs in a gas stream at a low flow rate [8].

But activated-carbon adsorption method needs to replace and dispose the adsorbent. The disposal of activated-carbon will cause the next environmental problem. Absorption is the process by which part of a gas mixture is transferred to a liquid based on the preferential solubility of the gas in the liquid.

This process is used most often to remove non-VOC gases, such as acid stack gases, but it is a complex and costly method of VOC control and removal. Adsorption is limited in its utility and not widely implemented in small industrial settings. Condensation is another non-destructive process for removal of VOCs from a gas stream. The VOC can be collected by cooling the gas stream to below the dew point of the VOC and then colleting the VOC in liquid form.

This process is relatively simple and most often used for low flow rate gas flows. But the cost of the process is high once the gas stream becomes complicated, which makes it economically inefficient.

Catalytic oxidation, similar to thermal incineration, is a combustion reaction between VOC and oxygen in the gas stream [9-10]. The difference between the two processes is that catalytic oxidation uses a catalyst coupled with temperature controls to facilitate the combustion reaction.

The limitation of catalytic oxidation is the certain VOCs will poison the catalyst and the cost of the process is a major concern in commercial operations. The catalytic combustion method requires a heat source, hence it is a highly energy-wasting process when VOC concentrations are low.

EXPERIMENT RESEARCH

3. Removal of HCHO with Photo-Catalytic Technology

In this section, the photo-catalytic degradation character of HCHO by TiO_2/UV, TiO_2/doped Ag/UV and TiO_2/doped Ce/UV was tested and compared. The effects of humidity, O_2, initial concentration, UV light intensity, and photo-catalyst amount on HCHO decomposition by TiO_2/UV were analyzed simultaneously. The byproduct was detected and the mechanism of HCHO degradation was performed.

3.1. Effect of doped Ag/TiO_2 or Ce/TiO_2 on Decomposition of HCHO

Catalyst is one of important factors for the pollutant decomposition. Different characteristic between TiO_2, Ag/TiO_2 and Ce/TiO_2 for HCHO decomposition was compared. Figure 1 illustrated the effect of doped Ag/Ce/TiO_2 on decomposition of HCHO. The conditions were as follows: flow-rate of 3 L/min, initial concentration of 0.1mg/m^3, relative humidity of 35%. It was found that conversions of HCHO in the TiO_2/UV, Ag-TiO_2/UV and Ce-TiO_2/UV processes were increased with irradiation time. It could be seen that the doping of silver or cerium ions could improve the photo-activity of TiO_2 effectively. Furthermore, the degradation character of the photo-catalyst was in the order Ce-TiO_2>Ag-TiO_2>TiO_2. The reason is as follows: silver or cerium species could create a charge imbalance, vacancies and unsaturated chemical bonds on the catalyst surface. It will lead to the increase of chemisorbed oxygen on the surface [11]. Surface chemisorbed oxygen has been reported [12] to be the most active oxygen, and plays an important role in oxidation reaction. Herein, redox reactions between the pollutant molecules and reactive oxygen radicals happened.

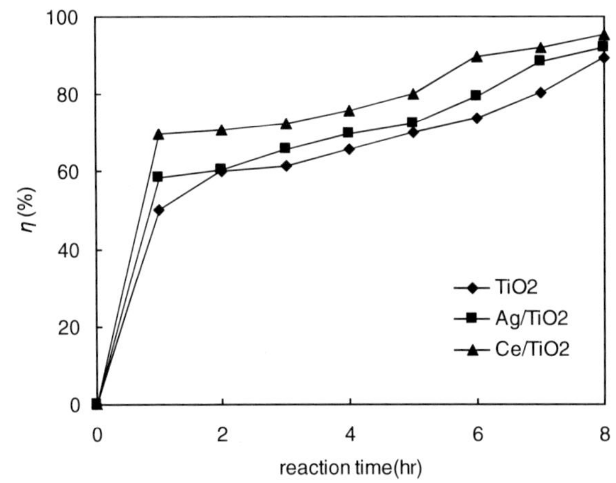

Figure 1. Effect of doped Ag/Ce/TiO_2 on decomposition of HCHO.

3.2. Effect of Relative Humidity of Air Stream

The effect of relative humidity (0-60% RH) of air stream on HCHO decomposition was examined by adding water vapor to a fixed concentration of HCHO. Figure 2 showed the experimental results at different relative humidity. The degradation rate increased with increasing relative humidity up to 35% and then started to decrease, which meant that 35% was the optimal humidity for photo-catalyst process under the experimental conditions. When the reaction time was 120min, the highest removal efficiency of HCHO was 60.2% when RH was 35%.

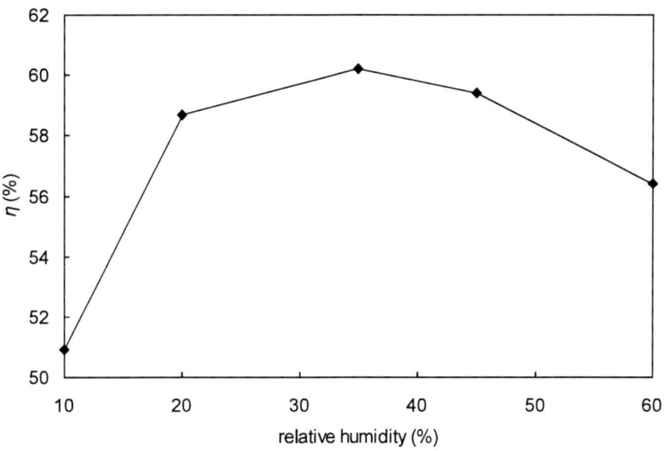

Figure 2. Effect of RH on decomposition of HCHO.

The enhancement of photo-catalytic reaction rate is frequently found in the presence of water vapor because hydroxyl groups or water molecules can behave as hole traps to form surface-adsorbed hydroxyl radicals [13-14]. In photo-catalyst process, the hydroxyl radicals formed on the illuminated TiO_2 can not only directly attack HCHO molecules, but also suppress the electron-hole recombination [15]. However, higher RH can be attributed to the competition for adsorption between HCHO and hydroxyl radicals, thus decrease the removal efficiency of HCHO.

3.3. Effect of O_2 Concentration

The effect of oxygen concentration on HCHO degradation was presented in Figure 3. The results corresponded to the initial concentration of 0.1mg/m^3, relative humidity of 35% and reaction time of 120min. It is obvious that oxidation rates for HCHO increased with increasing O_2 concentration under fixed conditions. As mentioned above, hydroxyl radical is an important factor to the HCHO photo-catalyst. At the same time, oxygen radical is also key factor for HCHO removal, which can react with HCHO on the TiO_2 surface and turn HCHO into CO_2 and H_2O. The similar conclusion was obtained by Aikaterini et al. [15].

3.4. Effect of Initial HCHO Concentration

The above results showed that HCHO could be removed with photo-catalytic processing. However, as a main indoor pollutant, the indoor HCHO concentration is usually below 0.5 ppmv [16]. It is worth discussing whether the low level of indoor HCHO can be decreased to a value below 0.1 mg/m^3 (specified in the indoor air quality standard of China).

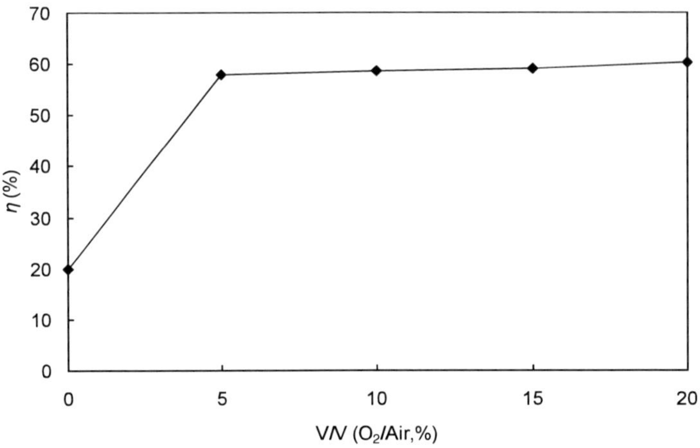

Figure 3. Effect of oxygen concentration on HCHO degradation.

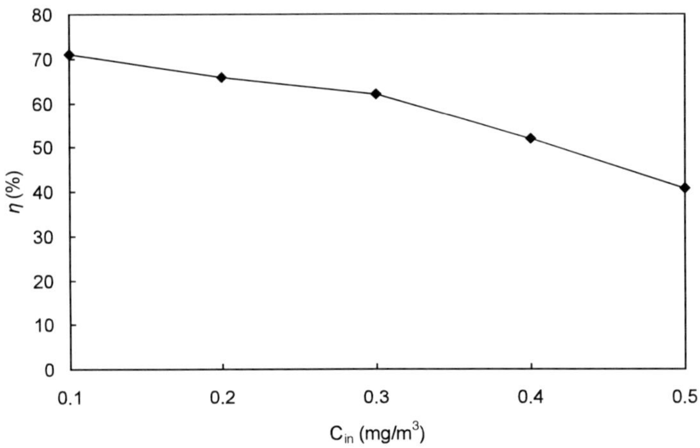

Figure 4. Effect of initial HCHO concentration on HCHO degradation.

So in our experiment, the HCHO concentrations in the experiment ranged between 0.1-0.5 mg/m^3.

The conditions were as follows: relative humidity of 35%, Ce-doped TiO_2 as photo-catalyst, and irradiation time of 120min. The results showed that the photo-catalytic degradation rates decreased with increasing HCHO initial concentration, just illustrated in Figure 4. In gas-phase photo-catalyst, collision frequency between radicals and HCHO affected the removal efficiency. When formaldehyde molecule reaches to the catalyst surface, the photo oxidation will occur. The number and energy of gas-phase reactive radicals (O•, OH•) in the reaction area did not change at a fixed UV lamp. Therefore, with constant gas flow rate, as initial HCHO concentration increases, unit HCHO obtains less energy, which results in a lower HCHO removal.

3.5. Effect of UV Light Wavelength

In order to investigate the influence of the UV intensity on the photo-catalytic efficiency, the experiments were performed using two lamp configuration (254 and 365 nm). TiO_2 photocatalyst was used in this experiment.

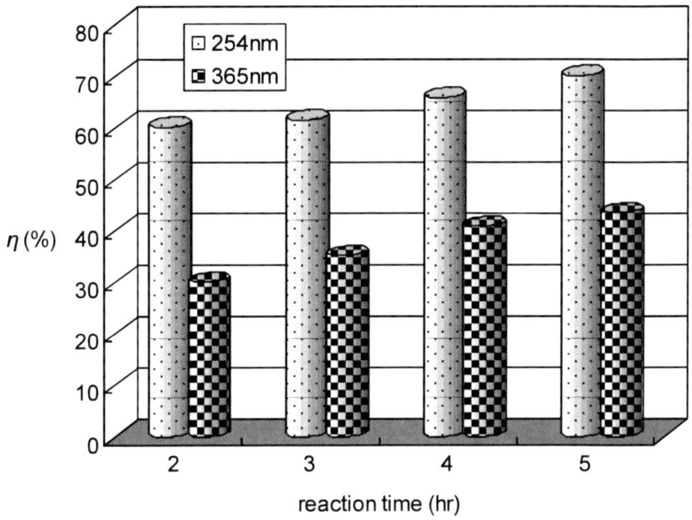

Figure 5. Effect of UV light wavelength on HCHO degradation.

The effect of UV light wavelength on the efficiency of HCHO degradation is shown in Fig 5. Just shown in Figure 5, 254 nm UV light provided more effective HCHO photo-degradation than 365 nm UV light. The different results obtained with 254 and 365 nm UV lamps were mainly due to the stronger UV irradiation from the 254 nm lamp (about 58 W/m^2 on its surface) than that from the 365 nm lamp (30 W/m^2 on its surface). This illustrates that the 254 nm UV lamp irradiated photons with higher energy, which led to more efficient degradation with TiO_2/UV.

3.6. Effect of the TiO_2 Amount

In photo-catalytic degradation of organic compounds, the optimal TiO_2 concentration depends mainly on both the nature of the compounds and the reactor geometry [17-18]. In this work, the influence of TiO_2 amount on HCHO photo degradation was investigated. A set of gaseous experiments with different amount of TiO_2 from 0 to 100mg was carried out at the RH of 35% and the initial HCHO concentration of 0.1mg/m^3. The degradation rates of HCHO for different amount TiO_2 were presented in Figure 6. The photo-catalytic degradation efficiency increased with increasing the amount of TiO_2 when TiO_2 amount was lower than 70mg. When the TiO_2 amount was more than 70mg, the photo-catalytic degradation efficiency was decreased. So 70mg of TiO_2 amount was the optional amount in our experiment.

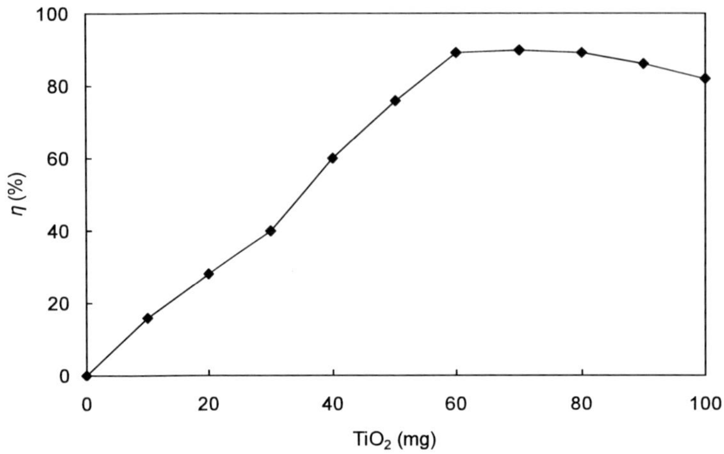

Figure 6. Effect of TiO_2 amount on HCHO degradation.

3.7. Mechanism of Photo-Catalytic Degradation of HCHO

The heterogeneous photo-catalytic process used in pollutant degradation involved the adsorption of pollutants on the surface sites, and the chemical reactions of converting pollutants into carbon dioxide and water.

Activation of TiO_2 is achieved through the absorption of a photon (hv) with ultra-band energy from UV irradiation source. This results in the promotion of an electron (e^-) from the valence band to the conduction band, with the generation of highly reactive positive holes (h^+) in the valence band. This caused aggressive oxidation of the surface-adsorbed toxic organic pollutants and converts them into CO_2 and water [19]. Published literatures showed that the most appropriate active sites for HCHO oxidation were surface noble metal sides [20].

During the HCHO decomposition by photo-catalytic processing, formic acid was identified as the intermediate from the photo-degradation of formaldehyde. In our experiment, ion chromatography (IC) was used to determine the byproducts by sampling the gas products into distilled water. The result in this study showed that formic acid was also found. The probably pathway of HCHO destruction was shown in Figure 7.

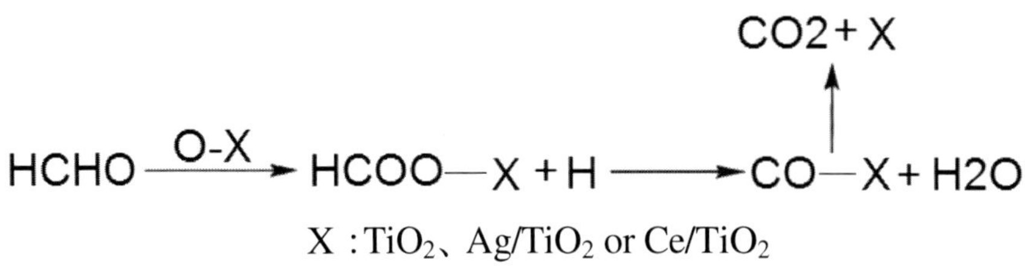

Figure 7. Suggested pathway for the photo-catalytic destruction of HCHO.

4. Removal of HCHO with Non-Thermal Plasma Technology

In this section, the difference among four kinds of $NaNO_2$ ferro-electric reactors was compared in terms of specific energy density (SED), energy yield (EY), HCHO decomposition. By-products during the decomposition of HCHO with non-thermal plasma technology and destruction mechanism were also investigated. In our experiment, three type of reactors whose fillings being dipped into 3000ppm, 5000ppm and 8000ppm $NaNO_2$ solution were represented by A, B and C, respectively.

4.1. Effect of $NaNO_2$ on HCHO Removal

Figure 8 represented the HCHO removal process in different $NaNO_2$ DBD reactor as a function of SED. The HCHO removal efficiency of different $NaNO_2$ reactors followed the same trend, that is, η_{HCHO} enhanced with increasing SED. Furthermore, HCHO removal enhanced with the increasing of $NaNO_2$ amount on ferro-electrics. For example, HCHO abatement efficiency of reactor A increased from 58% to 75% with increasing SED in the range of 220-680 J/l, while for reactor C, HCHO conversion was of 58%-93% with SED of 338-786J/l. And we also see that the conversion was almost the same for reactor B and C. Besides, $NaNO_2$ contributed to the increasing of SED andη_{HCHO}. From figure 8, we also can figure out that the η_{HCHO} of $NaNO_2$ reactors higher than the empty reactor. According to Baldur Eliasson et al. [21], the material of dielectric barrier plays a key role for the proper functioning of the discharge.

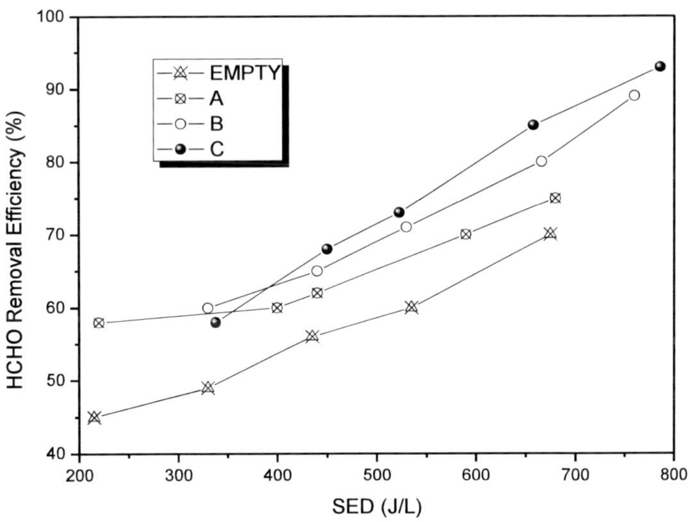

Figure 8. HCHO removal as a function of SED for $NaNO_2$ DBD reactor.

Van Durme et al. [22] indicated that the packing pellets are helpful for expanding the discharge region because the streamers (or microdischarges) are apt to propagate along the solid surface. From the experimental results, it demonstrated that the material employed in the experiment was an important factor influencing the HCHO removal. $NaNO_2$ in AC electric field could be polarized resulted in enhancing the electric field and more high-energy electrons produced that contributed to the abatement of HCHO. As the plasma discharge region expands, the concentrations of active species would increase due to the higher collision probability between electrons and gas molecules, leading to the acceleration of plasma chemistry reactions. Furthermore, from table 1, the BET surface area of different fillings was almost the same, which means that the main mechanism of HCHO destruction was plasma chemistry reaction.

Table 1. BET surface areas of the catalysts

Fillings	BET Surface Area(m^2/g)
ceramic raschig ring (none $NaNO_2$)	0.2451
$NaNO_2$ (3000ppm) coated on ceramic raschig ring	0.2852
$NaNO_2$ (5000ppm) coated on ceramic raschig ring	0.3058
$NaNO_2$ (8000ppm) coated on ceramic raschig ring	0.3172

In order to discuss the HCHO removal ability of the plasma reactor, energy yield was used. Figure 9 illustrated the relationship between EY and HCHO removal efficiency for different $NaNO_2$ DBD reactor. Obviously, with the HCHO removal efficiency enhancement, energy field descended for all of reactors. The highest of EY was 0.47 g/kWh for reactor A. Energy in plasma reactor was used to not only HCHO removal, but ferro-electric materials polarization, so EY decreased with the increasing of $NaNO_2$ amounts. Whereas the ferro-electric materials polarization could enhance the plasma discharge which increased the removal rate at the same time.

4.2. Effect of Amount of $NaNO_2$ of Ferro-Electrics Carrier on SED

Figure 10 showed the variation of SED as a function of applied voltage for various DBD reactors. $NaNO_2$ could increase the SED of DBD reactor effectively and the SED of three $NaNO_2$ reactors was all higher than that of the empty reactor. Furthermore, with increasing of applied voltage, SED of three kinds of $NaNO_2$ reactor increased. Through the experimental results, we could see that the amount of $NaNO_2$ on ferro-electrics was benefit for the improvement of the reactor performance effectively. Higher amount of $NaNO_2$ in the reactor was slightly superior to the smaller.

And the reactor B and C almost had the same SED when applied voltage was 9-13kV. When external AC voltage was applied across the high dielectric layer, $NaNO_2$ as ferro-electrics were polarized, and an intense electric field was formed around each pellet contact point, resulting in partial discharge. More amount of $NaNO_2$ led to more micro-discharges and the amount of active species in plasma.

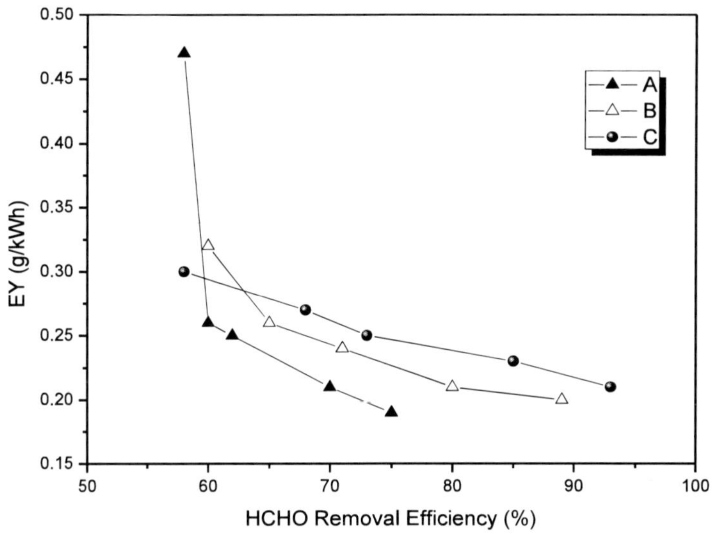

Figure 9. Relationship between EY and HCHO removal.

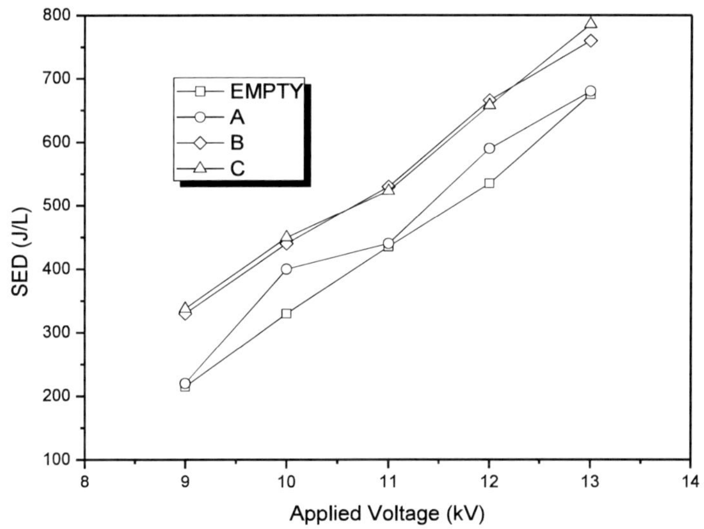

Figure 10. Variation of SED as a function of applied voltage for various $NaNO_2$ DBD reactors.

Taken the result of reactor A for example, with increasing the applied voltage from 10 kV to13 kV, the SED increased from 220 J/l to 680 J/l. As for reactor C, the SED increased from 338J/l to 786J/l under the same conditions.

4.3. HCHO Destruction by-Products and Mechanism Analysis

4.3.1. Ozone Generation in $NaNO_2$ Ferro-Electric Reactor

One of the most important by-products in a non-thermal plasma discharge was ozone.

Oxygen radicals are generated by dissociation of molecular oxygen after impact with accelerated electrons in the corona discharge. Atomic oxygen is a strong oxidizer, but its stability is very limited. Due to fast recombination processes, the lifetime is only a few microseconds at atmospheric pressure [23].

Oxygen radicals are generated by dissociation of molecular oxygen according to the reaction:

$$e + O_2 \rightarrow e + O + O$$

Atomic oxygen reacts with molecular oxygen in three-body collisions, forming ozone by the following reaction:

$$O + O_2 + M \rightarrow O_3 + M$$

where the third body M is needed to absorb the excess energy of the reaction. In air, M can be either molecular oxygen or molecular nitrogen [24]. Ozone is a more stable molecule; its lifetime at room temperature is several days. Ferro-electric materials affect the ozone production; Figure 11 showed an increase of ozone outlet production with increasing SED of three reactors. From mentioned above, due to the polarization and spontaneous polarization intensity of $NaNO_2$, we can concluded that the amounts of high energy electrons and atomic oxygen was of reactor C > reactor B > reactor A. So ozone outlet concentration of the reactors followed the same trends. The highest concentration for an energy density of 786 J/l was 100 mg/m^3 for reactor C, while the lowest concentration of 220 J/l was decreased to 65 mg/m^3 for reactor A.

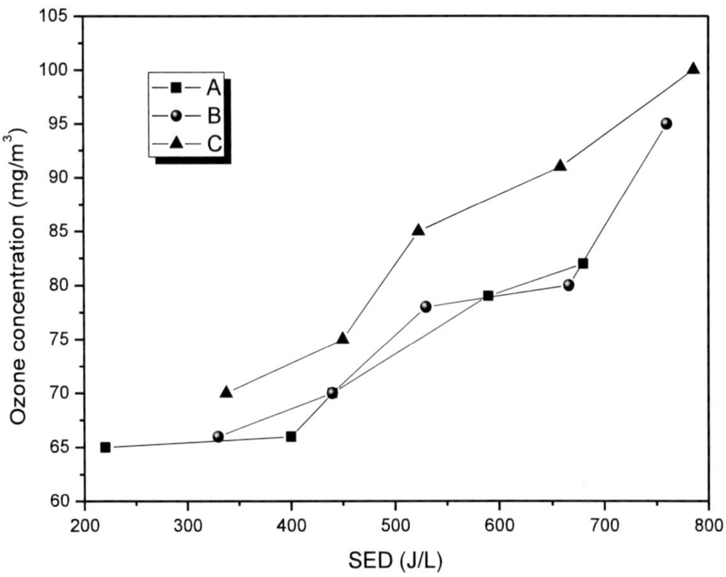

Figure 11. Ozone generation as a function of SED for $NaNO_2$ DBD reactor.

4.3.2. Destruction Mechanism for HCHO & Analysis for by-Products

The method of DBD plasma generates electrons with sufficient energy to cause the formation of gas-phase radicals, thereby driving the reactions of decomposition and oxidation of HCHO to form end products including H_2O, CO, and CO_2. According to the investigation of Moo Been Chang et al.[25], the removal of HCHO from gas streams depends on two removal mechanisms including (a) direct removal caused by the collision of electrons with HCHO (direct electron attack) and (b) reaction between HCHO molecules and gas-phase radicals (indirect gas-phase radical reaction). Gas-phase radicals may consist of OH, HO_2, H and O, which are very reactive and can react with HCHO to form other products with less environmental hazards. The reagent in our experiment was formaldehyde (analytical reagent) whose content was 98.5%. A little methanol was in the formaldehyde reagent. So in the chromatogram figure, a peak of methanol was detected. And from Figure 12 A and 12 B, we can only find out the peak of HCHO and CH_3OH. But the signal intensity of two peaks in Figure 12 B was lower than that in Figure 12 A, which meant HCHO and CH_3OH molecules were both destructed by plasma. Moreover, the products of the inlet and outlet were confirmed through GC-MS. From the mass spectrograms of inlet and outlet products, we could figure out that the main products of inlet gas were H_2O, N_2, HCHO and O_2, while in outlet gas, the main products were H_2O, N_2, less HCHO, O_2 and CO_2. CO (m/z=28.02) was possibly generated in this course, but the peaks of CO and N_2 probably overlapped each other, which needed to be further consideration.

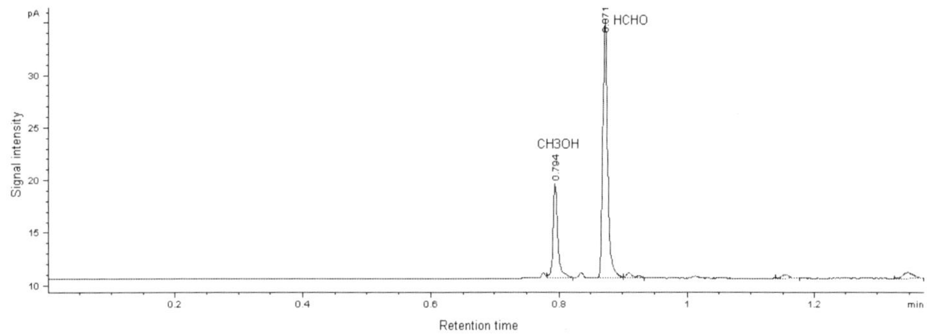

A Inlet

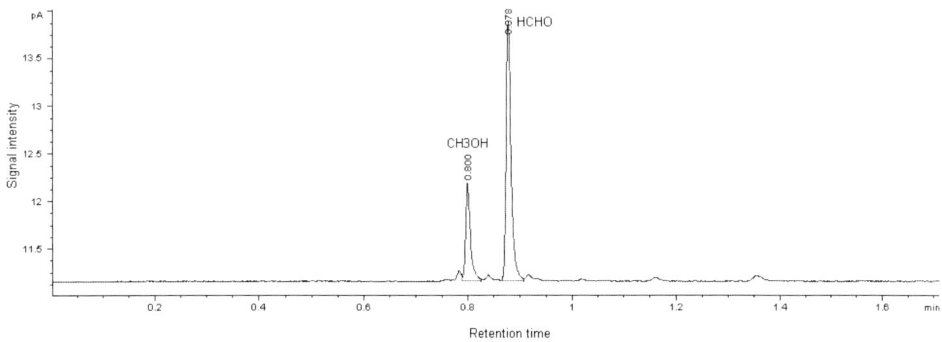

B Outlet

Figure 12. Products in inlet and outlet gases.

5. Removal of HCHO with Non-Thermal Plasma and Photocatalyst

In this section, three kinds of catalysts, that is, TiO_2, $BaTiO_3$, or TiO_2 & $BaTiO_3$ were tested for HCHO removal. Effect of catalyst carrier, catalysts, and catalyst amount on specific energy density (SED), energy yield (EY) and HCHO removal efficiency were investigated. The byproduct was detected and the mechanism of HCHO degradation was performed.

5.1. Appearance Characters of the Catalyst

Figure 13 showed the SEM photograph for TiO_2 & $BaTiO_3$ coated on catalyst carrier (a) ceramic Raschig rings, and (b) glass beads. From Figure 13, we could see that the particle shape was spherical, but the particles were coagulated each other. The surface-surface interactions between the Raschig rings or glass beads and catalyst particles allow easy deposition of catalyst particles. Besides, the catalyst coated on the ceramic Raschig rings was smooth than on the glass beads because ceramic Raschig rings had higher surface area, so ceramic Raschig rings could coat more catalysts than glass beads. The ceramic Raschig rings were used as catalyst carriers in our continuous experiment.

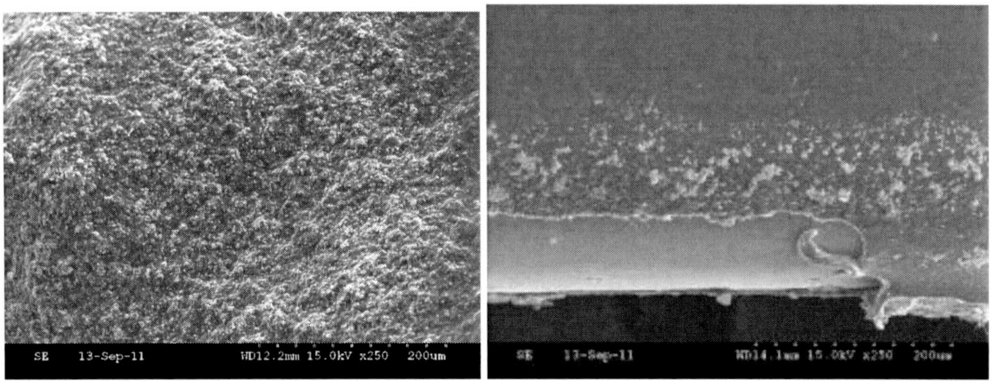

(a) (b)

Figure 13. SEM photographs of the TiO_2 & $BaTiO_3$ coated on catalyst carrier (a) ceramic Raschig rings (b) glass beads.

5.2. Electrical Discharge Characteristics

5.2.1. Effect of Catalyst Carrier on Plasma Discharge

Figure 14 shows the variation in SED as a function of applied voltage in the presence and absence of pellets. The SED values for reactors with different pellets were generally in the order: ceramic Raschig rings reactor > glass beads reactor > none pellets reactor under the same experimental conditions. At low applied voltage, SED increased slightly with applied voltage for all reactors. This order was maintained with further increase in voltage, as the SED increased substantially for all reactors, particularly those containing pellets. When the applied voltage was higher than 18 kV, the SED of the all reactors was inceased rapidly, especially for ceramic Raschig rings reactor and glass beads reactor. For example, as the applied voltage increased from 18 to 20 kV, the SED of the reactors increased from 45 J/l to 180 J/l (no pellets), 80 J/l to 260 J/l (glass beads), and 136 J/l to 315 J/l (ceramic Raschig rings).

Durme et al. [22] indicated that packing pellets were helpful for expanding the discharge region because the streamers (or micro-discharges) were apt to propagate along the solid surfaces. The relative dielectric constants of the ceramic and glass at normal temperature and pressure were 9.16 and 4.10, which meant that the energy storage ability of the ceramic was higher than that of glass. When the voltage was high enough, dielectric barrier discharge began to occur, and the pellets stored more energy as the electric field strength increased. More micro-discharges appeared under these electric fields, which resulted in more radicals and electrons.

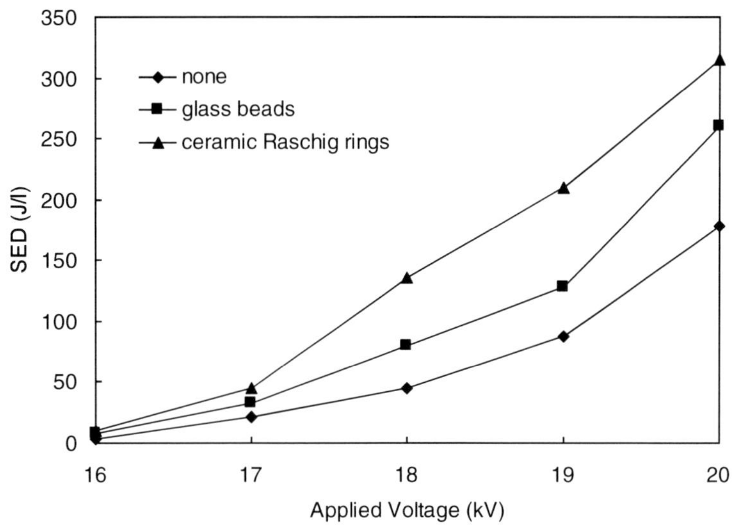

Figure 14. Variation in SED as a function of applied voltage for different catalyst carrier.

5.2.2. Effect of Catalysts on Plasma Discharge

Figure 15 shows the variation in SED as a function of applied voltage in the presence of different catalysts. The SED values for reactors with different catalysts were generally in the order: $BaTiO_3$+TiO_2 > $BaTiO_3$ > TiO_2 under the same experimental conditions. The mass ratio of $BaTiO_3$:TiO_2 was 3:1 and the amount of catalysts was 50 g. From Figure 15, we can see that SED increased with applied voltage for all catalysts. At low applied voltage, SED increased slightly with applied voltage for all reactors. For example, at an applied voltage of 16 kV, the SED for the three catalysts reactors were 10 J/l (TiO_2), 15 J/l ($BaTiO_3$) and 20 J/l ($BaTiO_3$+TiO_2). When applied voltage was higher than 17kV, the SED increased substantially for all reactors. Take applied voltage being 20 kV for example, the SED of the different catalysts reactors were 225 J/l (TiO_2), 360 J/l ($BaTiO_3$) and 458 J/l ($BaTiO_3$+TiO_2).

$BaTiO_3$, as a ferro-electric material, can be polarized in an electric field. When external AC voltage was applied across the high dielectric layer, $BaTiO_3$ were polarized, and an intense electric field was formed around each pellet contact point, resulting in partial discharge. As the applied ac voltage increased beyond the corona onset voltage, an increase of plasma activity was visible and the reactor was filled with high-energy free electrons [26].

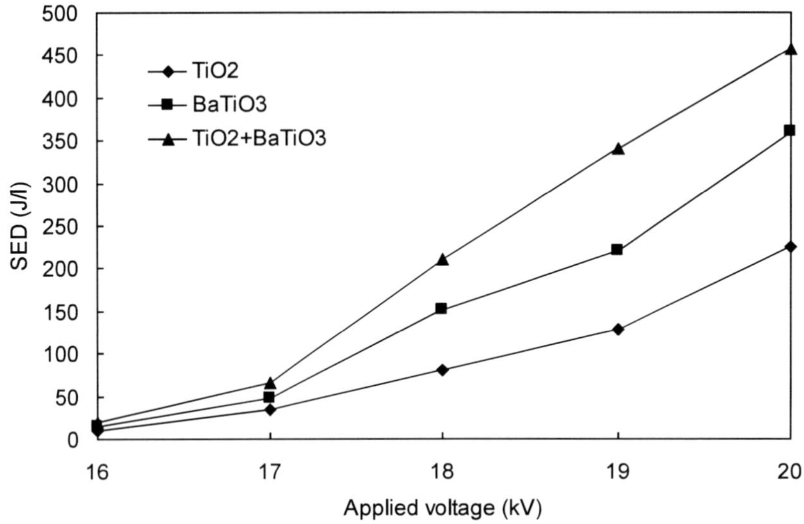

Figure 15. Variation of SED as a function of applied voltage for different catalyst.

Photo-catalysis has demonstrated to be very effective to treat pollutants under UV conditions. In our experiments, we tried to irrigate the TiO_2 using UV generated from plasma system. When applied voltage was low, the UV was weak and unable to irrigate the TiO_2. At the same time, $BaTiO_3$ could not be polarized and SED was low under this condition. When applied voltage was higher enough to generate more energetic radicals and electrons. And UV in the system increased also to irrigate the TiO_2 and lead to increase the SED of the system.

5.3. HCHO Decomposition in the Plasma

5.3.1. Effect of Catalyst Carriers on HCHO Removal

Figure 16 represented the HCHO removal process for reactors containing different catalyst carrier as a function of applied voltage. The HCHO removal efficiency was enhanced with increasing applied voltage. The removal efficiency of three reactors was in the order ceramic Raschig rings reactor > glass beads reactor > none reactor. For example, at 16 kV applied voltage, the efficiency values were: 60% (ceramic Raschig rings reactor), 50% (glass beads reactor) and 30% (none reactor), while at 20 kV applied voltage, the values were 82%, 75% and 65% respectively. Materials such as ceramics and glass were able to store energy and the energy storage ability of the ceramic was higher than that of glass because the relative dielectric constants of the ceramic was higher than the glass at normal temperature and pressure. When external AC voltage was applied across the materials layer and increased higher than ceramic or glass polarization, ceramic or glass materials were polarized, and the energy stored in the materials were emitted.

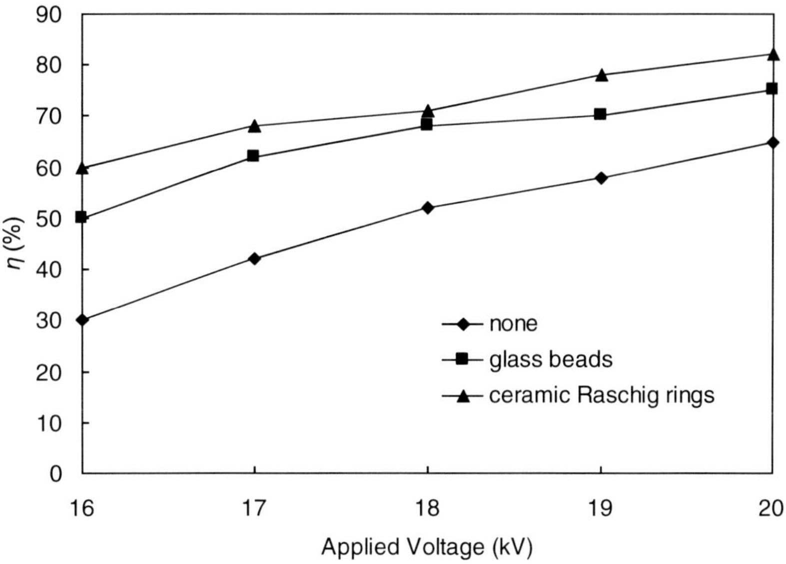

Figure 16. HCHO abatement as a function of applied voltage for different catalyst carrier reactors.

Thus more formaldehyde molecules, electrons and radicals could react with one another on the ceramic pellet surface than on the glass pellet surface. Furthermore, dielectric constant of ceramic is bigger than that of glass in same condition. Consequently, it could be concluded that stronger electric field is formed in ceramic Raschig rings packing case than in glass beads and that forms much more OH• radical to enhance the decomposition of HCHO.

5.3.2. Effect of Catalysts on HCHO Removal

It was proved that photocatalyst was put into plasma system could oxidize the VOCs effectively [27-29]. As we know, TiO_2 is a useful photocatalyst material and can be put into plasma system. It is controversial that whether UV light emitted from plasma can trigger TiO_2 and contribute to the decomposition of VOCs. Just as mentioned above, some researchers reported that plasma could emit plentiful UV light, which could lead to the photocatalysis of TiO_2 [30-31]; while some reported that the contribution of UV light from the plasma to activate photocatalyst was negligible [32-35]. In our study, spontaneous polarization material $BaTiO_3$ and photocatalyst TiO_2 was put into the plasma system to investigate the removal of HCHO. In our previous studies, ferroelectric materials such as $BaTiO_3$ had been proved that it could enhance the electric field [36-37]. When external AC voltage was applied across the high dielectric layer, $BaTiO_3$ were polarized, and an intense electric field was formed around each pellet contact point, resulting in partial discharge. As the applied ac voltage increased beyond the corona onset voltage, an increase of plasma activity was visible and the reactor was filled with high-energy free electrons [26].

Figure 17 represented the removal process of HCHO in different catalysts reactor. The mass ratio of $BaTiO_3$:TiO_2 was 3:1 and the amount of catalysts was 50 g. The HCHO conversion of three kinds of catalyst was TiO_2 & $BaTiO_3$ > $BaTiO_3$ > TiO_2. From 16kV to 20kV, HCHO abatement efficiency followed the same trends. For example, η_{HCHO} of TiO_2 & $BaTiO_3$ increased from 65% to 93% with increasing applied voltage in the range of 16-20kV, while η_{HCHO} was of 58%-82% for $BaTiO_3$ and of 48%-75% for TiO_2. With increasing the applied voltage, energy throughout the entire discharge volume was developing; micro-discharges occurred in the whole reactor in turn and stimulated the chemical reactions between HCHO molecules and active electrons. $BaTiO_3$ in the plasma system could enhance the electric field and UV light, which could lead to the generation of more active radicals such as OH• and increased the removal efficiency of HCHO. Furthermore, ozone generated in plasma system could be decomposed by adding UV light to an active oxygen atom. The active oxygen atom is a strong oxidant for HCHO. The improvement of removal efficiency of HCHO might also be due to the numerous production of OH• which is a very strong oxidant.

5.3.3. Effect of Amount of TiO_2 & $BaTiO_3$ Catalyst on HCHO Removal

Figure 18 showed the variation of η_{HCHO} as a function of applied voltage for the catalyst. Four types of catalysts (amount ratio of $BaTiO_3$ & TiO_2 was 3:1, 2:1, 1:1,1:2) were used for the HCHO removal. From Figure 18, we could see that the HCHO removal efficiency was highest when $BaTiO_3$ & TiO_2 amount ratio was 3:1. The HCHO conversion of four kinds of catalyst was $BaTiO_3$ & TiO_2 amount ratio being 3:1 > 2:1 > 1:1 > 1:2. BET surface area of the different catalysts was tested (see Table 2). The results showed that BET surface area decreased with increasing TiO_2 amount in $BaTiO_3$ & TiO_2. From experimental results, we can conclude that higher amount of $BaTiO_3$ in $BaTiO_3$ & TiO_2 was slightly superior to the smaller.

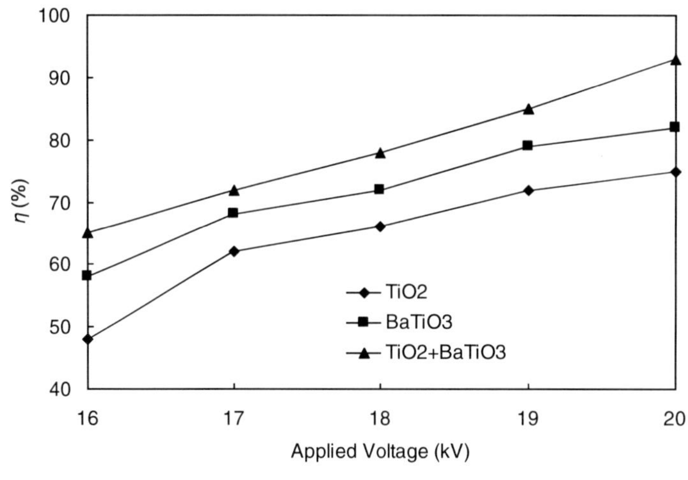

Figure 17. HCHO abatement as a function of applied voltage for different catalysts reactors.

Taken the result of $BaTiO_3$ & TiO_2 being 3:1 for example, with increasing the applied voltage from 16 kV to 20 kV, η_{HCHO} increased from 68% to 93%. As for $BaTiO_3$ & TiO_2 being 1:2, η_{HCHO} increased from 41% to 68% under the same conditions. Just mentioned above, $BaTiO_3$ could enhance the electric field. In a higher electric field, an increase of plasma activity was visible and the reactor was filled with high-energy free electrons, which also led to the increase of UV light. So higher amount of $BaTiO_3$ in $BaTiO_3$ & TiO_2 was benefit to the removal of HCHO.

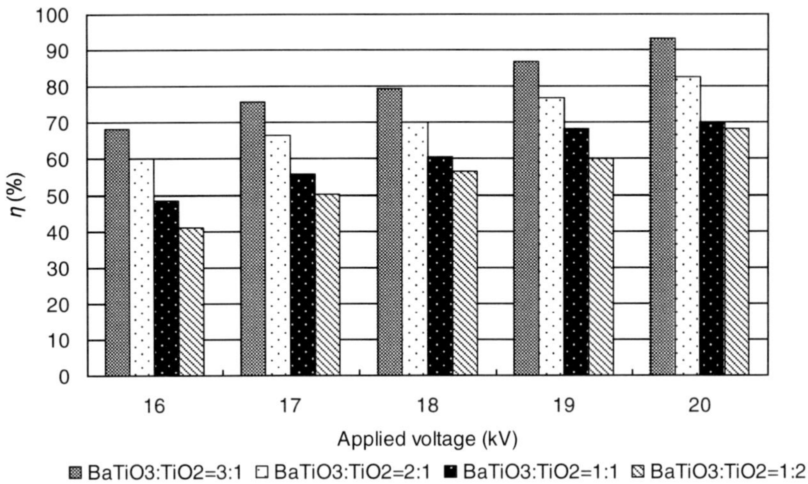

Figure 18. HCHO abatement as a function of applied voltage for different catalysts.

Table 2. BET surface area for different catalysts

Catalysts		BET (m^2/g)
$BaTiO_3/TiO_2$	$mBaTiO_3$: $mTiO_2$ = 3:1	18.4
	2:1	15.3
	1:1	12.4
	1:2	10.2

5.3.4. Relationship between HCHO Removal Efficiency and EY

In order to discuss the HCHO removal ability of the plasma reactor, energy yield was used. Figure 19 illustrated the relationship between EY and HCHO removal efficiency for different catalysts reactor. Obviously, with the HCHO removal efficiency enhancement, energy field descended for all of reactors. Energy in plasma reactor was used to not only HCHO removal, but ferroelectric materials polarization, so EY decreased with the HCHO removal efficiency Whereas the ferroelectric materials polarization could enhance the plasma discharge which increased the removal rate at the same time. Furthermore, the energy yield is in the order of TiO_2 & $BaTiO_3$ > $BaTiO_3$ > TiO_2.

5.4. Destruction Mechanism for HCHO and Analysis for by-Products

A DBD plasma generates electrons with sufficient energy to cause the formation of gas-phase radicals, thereby driving the reactions of decomposition and oxidation of HCHO to form end products including H_2O, CO and CO_2. The removal of HCHO probably depends on two mechanisms: (a) direct removal caused by the collision of electrons and (b) reaction between HCHO molecules and gas-phase radicals (indirect gas-phase radical reaction). Gas-phase radicals may consist of O•, OH•, HO_2• and O_3.

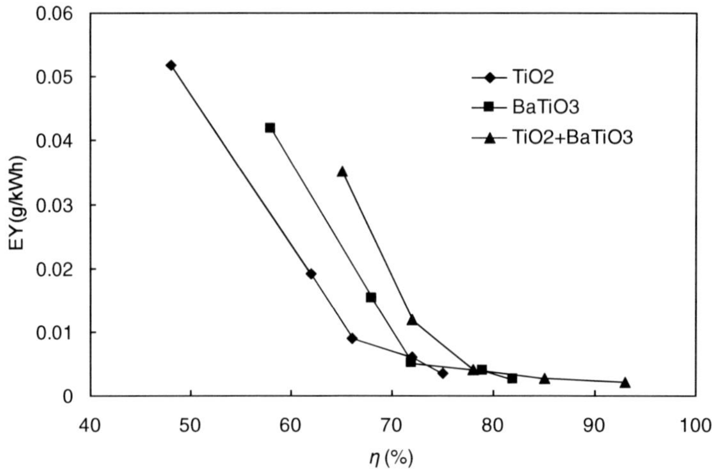

Figure 19. Energy yield as a function of HCHO removal efficiency for catalysts.

The main purpose of the paper was to discuss the by-products in the system. The reagent in our experiment was formaldehyde (analytical reagent) whose content was 98.5%.

A little methanol was in the formaldehyde reagent. So in the chromatogram figure, a peak of methanol was detected. And from Figure 20, we can only find out the peak of HCHO and CH_3OH. But the signal intensity of two peaks in inlet effluent was lower than that in outlet effluent, which meant HCHO and CH_3OH molecules were both destructed by plasma.

During the decomposition of HCHO, the carbon balance and CO_2 selectivity are useful parameters to assess the performance of the catalyst in the plasma reactor.

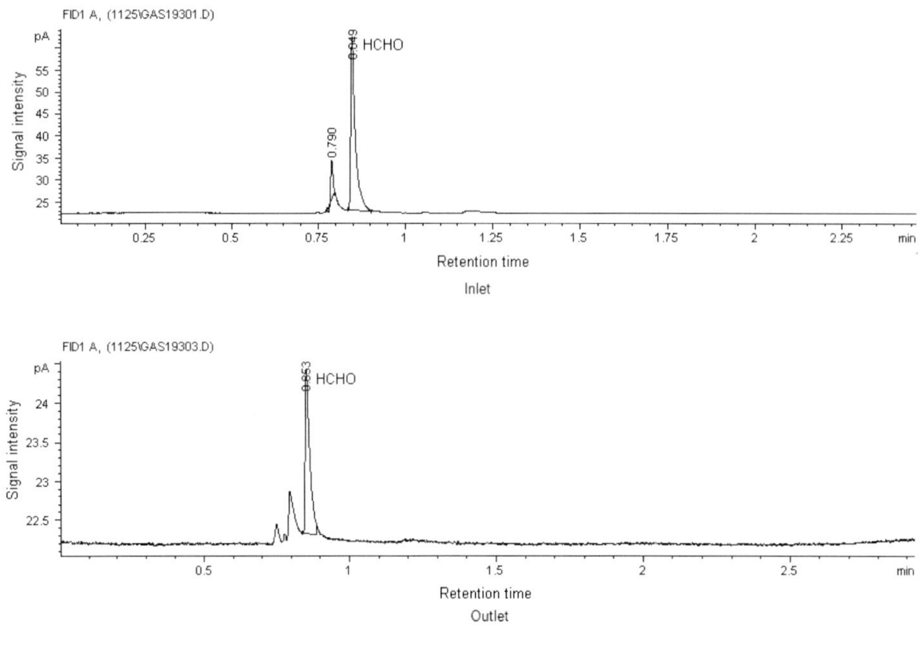

Figure 20. Products in inlet and outlet gases.

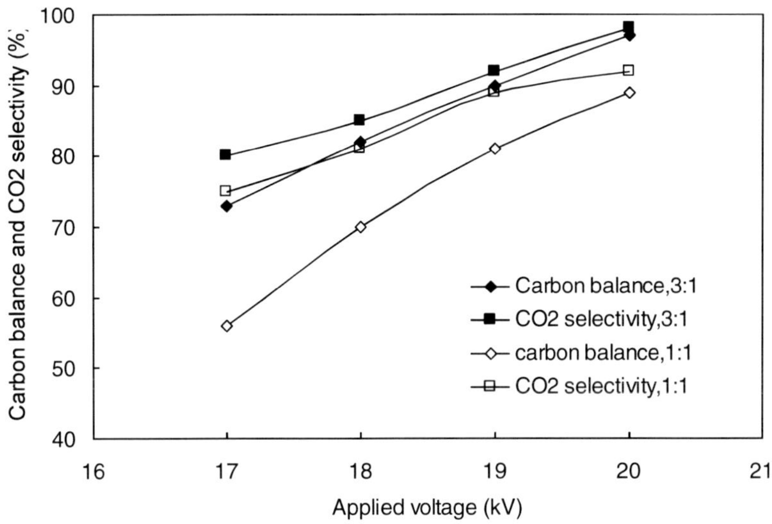

Figure 21. Carbon balance and CO_2 selectivity for HCHO decomposition.

In Figure 21, quantitative analysis data concerning carbon balance and CO_2 selectivity for two kinds of catalysts in the plasma reactor at different applied voltage were presented. From Figure 21, it was obviously that carbon balance and CO_2 selectivity consistently increased with applied voltage, indicating that less intermediates formed during plasma process for HCHO. Take $BaTiO_3$ & TiO_2 catalyst being 3:1 for example, when applied voltage was 20 kV, carbon balance and CO_2 selectivity were 97% and 98%, respectively.

CONCLUSION

1. Nano-structured TiO_2, Ag-TiO_2 and Ce-TiO_2 thin films were prepared by sol-gel method at room temperature. Furthermore, the TiO_2 thin film was used in the photo-catalyst reactor for HCHO decomposition. The effects of doped Ag/Ce/TiO_2, relative humidity, oxygen concentration, initial HCHO concentration, UV light wavelength and TiO_2 amount on HCHO decomposition were investigated. The results showed that doped Ag or Ce ions could enhance the photo-catalyst ability. The degradation character of the photo-catalyst was in the order Ce-TiO_2>Ag-TiO_2>TiO_2. 35% was the optimal humidity for photo-catalyst process under the experimental conditions. Higher concentration of oxygen was better for HCHO removal. Photo-catalytic degradation rates decreased with increasing HCHO initial concentration. Illumination using a 254 nm light source was better than 365 nm. The photo-catalytic degradation efficiency increased with increasing the amount of TiO_2 when TiO_2 amount was lower than 70mg. Formic acid was the main byproduct during the decomposition of HCHO. The reaction rate constant (k) of TiO_2、Ag/TiO_2、Ce/TiO_2 was sequenced $k_{Ce/TiO2}>k_{Ag/TiO2}>k_{TiO2}$, meaning that Ce/$TiO_2$ had the best photo-catalytic abilities among the catalysts.

2. The abatement of HCHO via $NaNO_2$ ferro-electric packed bed dielectric barrier discharge plasma in a coaxial cylindrical reactor was experimentally investigated. The main

results showed that the SED depended almost linearly on the voltage for $NaNO_2$ ferro-electric DBD reactor. HCHO removal efficiency enhanced with increasing SED for all $NaNO_2$ reactors. More amount of $NaNO_2$ contributed to higher HCHO removal efficiency and ozone generation in the reactors. As an important by-product, ozone concentration increased with higher SED. The possible main products in the outlet effluent were CO, CO_2 and H_2O.

3. Destruction of indoor air pollutant HCHO using spontaneous polarization plasma and photocatalyst was carried out at atmospheric pressure and room temperature. The number of microdischarge pulses of different catalysts is in the order of TiO_2 & $BaTiO_3$ > $BaTiO_3$ > TiO_2. The SED values for reactors with different catalysts were generally in the order: TiO_2 & $BaTiO_3$ > $BaTiO_3$ > TiO_2 under the same experimental conditions. The HCHO removal efficiencies followed the same trend, that is, was enhanced with increasing applied voltage. The removal efficiency of three reactors was in the order ceramic Raschig rings reactor > glass beads reactor > none reactor for the same experimental conditions. The HCHO conversion of three kinds of catalyst was TiO_2 & $BaTiO_3$ > $BaTiO_3$ > TiO_2. HCHO removal efficiency was highest when $BaTiO_3$ & TiO_2 $BaTiO_3$ amount ratio was 3:1. Higher amount of $BaTiO_3$ in TiO_2 & $BaTiO_3$ was slightly superior to the smaller. As the applied voltage increased, the energy yield increased for all catalysts. The energy yield is in the order of TiO_2 & $BaTiO_3$ > $BaTiO_3$ > TiO_2. Carbon balance and CO_2 selectivity for all catalysts consistently increased with applied voltage, indicating that less intermediates formed during plasma process for HCHO.

ACKNOWLEDGMENTS

This work was supported by the National High-technology Research and Development Program of China (2011AA03A406) and National Natural Science Foundation of China (21307003).

REFERENCES

[1] IARC. (2006) Formaldehyde, 2-butoxyethanol and 1-tert-butoxypropan-2-ol. Monographs on the evaluation of carcinogenic risks to humans. LYON, France.

[2] NICNAS. (2006) National Industrial Chemicals Notification and Assessment Scheme: *Formaldehyde, Priority Existing Chemical Assessment Report*. Australian Government.

[3] Yue, W. Jin, X.B. Pan, X.C. Ding, J. (2004) Relationship between indoor air formaldehyde exposure and allergic asthma in adults. *Chin. J. Public Health*. 20, 904-906.

[4] Edling, C. Hellquist, H. Odkvist, L. (1988) Occupational exposure to formaldehyde and histopathological changes in the nasal mucosa. *Br. J. Ind. Med.* 45, 761-765.

[5] Wilhelmsson, B. Holmstrom, M. (1992) Possible mechanisms of formaldehyde-induced discomfort in the upper airways. *Scand J Work Environ Health*. 18, 403-407.

[6] Zhang, L. Steinmaus, C. Eastmond, D.A. Xin, X.K. Smith, M.T. (2009) Formaldehyde exposure and leukemia: a newmeta-analysis and potentialmechanisms. *Mutat. Res*. 681, 150-168.

[7] McGregor, D. Bolt, H. Cogliano, V. Richter-Reichhelm, H.B. (2006) Formaldehyde and glutaraldehyde and nasal cytotoxicity: case study within the context of the 2006 IPCS Human Framework for the Analysis of a cancer mode of action for humans. Crit Rev Toxicol, 36, 821-835.

[8] Spivey, J.J. (1988) Recovery of volatile organics from small industrial sources. *Environmental Progress* 7, 31-40.

[9] van der Vaart, D.R. Spivey, J.J. Vatavuk, W.M. Wehe, A. (1990) EPA 450/3-90-006 US EPA/OAQPS, Research Triangle Park, NC.

[10] Spivey, J.J. (1990) Innovative Hazardous Waste Treatment Technologies. H. Freeman, ed., Technomic Publishing,

[11] Yang, S.X. Zhu, W.P. Jiang, Z.P. Chen, Z.X. Wang, J.B. (2006) The surface properties and the activities in catalytic wet air oxidation over CeO_2–TiO_2 catalysts. *Appl. Surf. Sci.* 252, 8499-8505.

[12] Jing, L.Q. Xu, Z.L. Sun, X.J. Shang, J. Cai, W.M. (2001) The surface properties and photocatalytic activities of ZnO ultrafine particles. *Appl. Surf. Sci.* 180,308-314.

[13] Jacoby, W.A. Blake, D.M. Noble, R.D. Koval, C.A. (1995) Kinetics of the Oxidation of Trichloroethylene in Air via Heterogeneous Photocatalysis. *J. Catal.* 157, 87-96.

[14] Kim, S.B. Hong, S.C. (2002) Kinetic study for photocatalytic degradation of volatile organic compounds in air using thin film TiO_2 photocatalyst. *Appl. Catal. B* 35, 305-315.

[15] Boulamanti, A.K. Philippopoulos, C.J. (2008) Photocatalytic degradation of methyl tert-butyl ether in the gas-phase: A kinetic study. *J. Hazard. Mater.* 160, 83-87.

[16] Lu, Y.W. Wang, D.H. Ma, C.F. Yang, H.C. (2010) The effect of activated carbon adsorption on the photocatalytic removal of formaldehyde. *Building and Environment* 45, 615-621.

[17] Shiraishi, F. Yamaguchi, S. Ohbuchi, Y. (2003) A rapid treatment of formaldehyde in a highly tight room using a photocatalytic reactor combined with a continuous adsorption and desorption apparatus. *Chem. Eng. Sci.* 58, 929-934.

[18] Liu, H.M. Lian, Z.W. Ye, X.J. Shangguan, W.F. (2005) Kinetic analysis of photocatalytic oxidation of gas-phase formaldehyde over titanium dioxide. *Chemosphere* 60, 630-635.

[19] Zou, L.D. Luo, Y.G. Hooper, M. Hu, E. (2006) Removal of VOCs by photocatalysis process using adsorption enhanced TiO_2–SiO_2 catalyst. *Chem. Eng. Process.* 45, 959-964.

[20] Kecske´s T., Rasko´J., Kiss J. (2004) FTIR and mass spectrometric studies on the interaction of formaldehyde with TiO_2 supported Pt and Au catalysts. *Appl. Catal. A* 273, 55-62.

[21] Eliasson, B. Kogelschatz, U. (1991) Nonequilibrium volume plasma chemical processing, *IEEE Trans. Plasma Sci.* 19, 1063-1077.

[22] Durme, J.V. Dewulf, J. Leys, C. Langenhove, H. (2008) Combining non-thermal plasma with heterogeneous catalysis in waste gas treatment: A review, *Appl. Catal., B* 78, 324-333.

[23] Oda, T. Yamashita, Y. Takezawa, K. Ono, R. (2005) Oxygen atom behaviour in the nonthermal plasma, *Thin Solid Films* 506/507, 669–673.

[24] Magureanu, M. Mandache, N.B. Eloy, P. Gaigneaux, E.M. Parvulescu, V.I. (2005) Plasma-assisted catalysis for volatile organic compounds abatement, *Appl. Catal. B:Environ.* 61, 12–20.

[25] Chang M.B., Lee C.C. (1995) Destruction of formaldehyde with dielectric barrier discharge plasmas, *Environ. Sci. Technol.* 29, 181-186.

[26] Yamamoto, T., Pamanathan, K., Lawless, P.A., Ensor, D.S., Newsome, J.R., Plaks, N. and Ramsey, G.H. (1992) Control of Volatile Organic Compounds by an ac Energized Ferroelectric Pellet Reactor and a Pulsed Corona Reactor. *IEEE Trans. Ind. Applicat.* 28, 528-534.

[27] Ao, C.H., Lee, S.C., Yu, J.Z. and Xu, J.H. (2004) Photodegradation of formaldehyde by photocatalyst TiO_2: effects on the presences of NO, SO_2 and VOCs. *Appl. Catal. B* 54, 41-50.

[28] Zhang, C.B. and He, H. (2007) A comparative study of TiO_2 supported noble metal catalysts for the oxidation of formaldehyde at room temperature. *Catal. Today* 126, 345-350.

[29] Zhang, C.B., He, H. and Tanaka, K.I. (2006) Catalytic performance and mechanism of a Pt/TiO_2 catalyst for the oxidation of formaldehyde at room temperature. *Appl. Catal.* B 65, 37-43.

[30] Zhu,T., Wan,Y.D., He,X.W., Xu, D.Y., Shu, X.Q.,Fang, Y. and Chen, X. (2011) Effect of Modified Ferroelectric on Nonthermal Plasma Process for Toluene Decomposition. Fresen. *Environ. Bull.* 20, 149-155.

[31] Ye, Z.L., Luo, S.L., Xiong, J.Y. and Ma, F. (2010) Decomposition of Isobutyl Acetate Waste Gas with Combined Plasma Photolysis Technology. *Fresen. Environ. Bull.* 19, 2490-2496.

[32] Ogata, A., Kim, H.H., Futamura, S., Kushiyama, S. and Mizuno, K. (2004) Effects of catalysts and additives on fluorocarbon removal with surface discharge plasma. *Appl Catal B: Environ.* 53, 175-180.

[33] Sano, T., Negishi, N., Sakai, E. and Matsuzawa, S. (2006) Contributions of photocatalytic/catalytic activities of TiO_2 and γ-Al_2O_3 in nonthermal plasma on oxidation of acetaldehyde and CO. *J. Mol. Catal. A: Chem.* 245, 235-241.

[34] Huang, H.B., Ye, D.Q., Fu, M.L. and Feng, F.D. (2007) Contribution of UV light to the decomposition of toluene in dielectric barrier discharge plasma/photocatalysis system. *Plasma Chem. Plasma Process* 27, 577-588.

[35] Huang, H.B. and Ye, D.Q. (2009) Combination of photocatalysis downstream the non-thermal plasma reactor for oxidation of gas-phase toluene. *J. Hazard. Mater.* 171, 535-541.

[36] Liang, W.J., Li, J., Li, J. and Jin, Y.Q. (2009) Abatement of toluene from gas streams via ferro-electric packed bed dielectric barrier discharge plasma. *J. Hazard. Mater.* 170, 633-638.

[37] Liang, W.J., Li, J., Li, J.X., Zhu, T. and Jin, Y.Q. (2010) Formaldehyde removal from gas streams by means of $NaNO_2$ dielectric barrier discharge plasma. *J. Hazard. Mater.* 175, 1090-1095.

In: Formaldehyde
Editor: Amy Patton

ISBN: 978-1-63482-412-5

Chapter 6

FORMALDEHYDE: DETECTION AND BIODEGRADATION

Galina Gayda*[*,1], Olha Demkiv[1], Halyna Klepach[1,2], Mykhailo Gonchar[1,3], Revital Levy-Halaf[4], David Wolf[4] and Marina Nisnevitch[4]

[1]Department of Analytical Biotechnology, Institute of Cell Biology, Lviv, Ukraine
[2]Drohobych Ivan Franko State Pedagogical University, Drohobych, Ukraine
[3]Institute of Applied Biotechnology and Basic Sciences, University of Rzeszow, Kolbuszowa, Poland
[4]Department of Chemical Engineering, Biotechnology and Materials, Ariel University, Ariel, Israel

ABSTRACT

Formaldehyde (FA) is a highly reactive compound with a toxic effect on all organisms due to non-specific interactions with proteins and nucleic acids. This substance is a hazardous air pollutant, and prolonged exposure to FA may cause serious health effects. This chapter describes approaches for detection and eliminating FA in several real and model sources.

The cells of the thermotolerant methylotrophic yeast *Hansenula (Ogatae) polymorpha* as well as FA-oxidizing enzymes isolated from these cells were employed as catalytic elements in the development of enzymatic methods for FA detection by analytical kits and biosensors as well as for degradation in bioreactors. Recombinant yeast strains, originating from NCYC 495 and CBS 4732 *H. polymorpha* strains resistent to elevated concentrations of FA in the medium and overproducing a homologous NAD- and glutathione-dependent formaldehyde dehydrogenase (FdDH), were constructed. Optimal cultivation conditions for the highest yield of FdDH as well as formaldehyde reductase (FR) were found. Alcohol oxidase (AOX) was isolated from the mutant strain

[*] Galina Gayda E-mail: galina.gayda@gmail.com.

H. polymorpha C-105 (*gcr1 catX*) which is defective in glucose repression of AOX synthesis and lacks catalase. Simple schemes for FdDH, FR and AOX isolation and chromatographic purification from the recombinant overproducers were proposed and the physico-chemical and catalytic properties of the purified enzymes were studied. Enzymatic methods and analytical kits for FA assay based on glutathione-dependent FdDH and AOX were developed. The purified FdDH and AOX preparations, as well as yeast cells overproducing these enzymes, were used for the construction of FA-selective enzyme-based and microbial electrochemical biosensors.

A laboratory prototype of a FA analyzer with a catalytic chemosensing element and a simple low-cost FA-sensitive biosensor for semi-quantitative FA detection were developed. The latter sensor is based on a change in the color of a solution that contains a mixture of AOX, horseradish peroxidase and a chromogen, after exposure to airborne FA. The reliability of the proposed analytical approaches was tested on model solutions and on real samples of FA-containing industrial products, vaccines and wastes. A strong cross-correlation was found between FA content values obtained by the developed methods and routinely used chemical ones.

The enzyme preparations and the yeast cells overproducing the corresponding enzymes were used for construction of FA-selective electrochemical biosensors. The purified enzymes and the recombinant and mutant cells served as the catalytic units in cartridges for FA removal from indoor air and from water solutions.

Experimental data confirm the possibility of exploiting the developed bioreactors based on crude preparations of AOX or methylotrophic yeast cells, for effective FA oxidation (up to 99.9%) controlled by a coupled FdDH-based biosensor.

Keywords: formaldehyde, biodegradation, enzymatic assay, biosensor, chemosensor, air-lift bioreactor

INTRODUCTION

Formaldehyde (FA) is an extremely toxic agent which is found in over 2000 commercial products. In 2013 its annual production was estimated to be more than 40 million metric tonnes (Magnusson, 2013). FA has been used in many industries and consumer products as an antibacterial agent, antiseptic material, fumigant, bactericide, fungicide and insecticide. Use of FA in the food industry, where it serves as an antibacterial agent and as a preservative during food processing, is particularly important. Many industrial enterprises use it as a key chemical compound in organic synthesis and in combination with phenol for commercial production of composites, synthetic resins (urea-formaldehyde and phenol-formaldehyde, pentaerythritol and other compounds), adhesives, plastics, as a structural material in the automobile industry and for electrical engineering (Magnusson, 2013). FA is also used in the textile industry as a component of a coupling agent that ensures the stability of fabric shapes. Wastewaters from many different industries contain FA and phenols, which are hazardous to living organisms even at low concentrations. Fine cleaning and toxic substances control are therefore essential prior to the discharge of such wastewaters into the environment.

However, FA is also a natural metabolite of all living organisms. It has been found in fruits, vegetables, flesh, and biological fluids of human origin (Gerberich and Seaman, 1994, 2007). In extreme cases, some frozen fish, especially from the *Gadoid* species, can accumulate up to 200 mg of FA per kilogram wet weight due to enzymatic degradation of a natural fish component - trimethylamine oxide (Rehbein et al., 1995; Pavlishko et al., 2003). FA is also emitted by bacteria, algae and plankton (Liteplo et al., 2002).

Humans are exposed to FA from a variety of sources. The major source of atmospheric FA is related to combustion processes: forest fires, automobile exhaust and photooxidation of hydrocarbons. In water, FA is also formed by the irradiation of humus substances by sunlight (Liteplo et al., 2002; National Research Council, 1982). Additional exposure to FA emissions comes from its use as an embalming fluid in anatomy labs, morgues, and from its use as a fumigant and sterilant (Liteplo et al., 2002).

According to data of the International Agency for Research on Cancer (IARC) (IARC, 1995 and 2006; Squire and Cameron, 1984; Heck et al., 1985; Feron et al., 1991; Usanmaz et al., 2002), FA is a toxic compound with mutagenic, immunogenic, allergenic and carcinogenic effects. FA toxicity is the result of its very high activity: FA easily reacts with proteins and nucleic acids, causing their post-synthetic modifications and disruption of natural functions (Conaway et al., 1996; Quievrin and Zhitkovich, 2000; Tokarchuk and Zaitseva, 2014). FA has recently been described as one of the chemical mediators of apoptosis (Monticello et al., 1996; Yu, 1998; Yu et al., 2003; Szende et al., 1995). Interest in the possible role of FA in the pathogenesis of diabetic angiopathies, atherosclerosis and cardiovascular diseases has also increased in recent years (Dalbey, 1982; Kerns et al., 1983; Hayes et al., 1986; Hansen et al., 1995; Hauptmann et al., 2004).

FA is currently considered to be the main cause of the sick building syndrome, which is defined as a set of symptoms associated with irritation of the upper respiratory tract and the eyes caused by harmful compounds (particularly FA) which are found in building materials, tobacco smoke, some medicinal preparations, etc. (Dingle, P., 1997; LARCIARC, 1995; Kim et al., 2002). FA is a widespread chemical pollutant of water, air and soil. It can be found in food as well as in industrial and domestic sewage.

FA in indoor air causes several health problems. Exposure to airborne FA at 1-3 ppm leads to irritation of the eyes, nose and throat, and exposure to 10-20 ppm results in eye irritation and a burning sensation in the nose and throat as well as breathing difficulties that lead to coughing. However, many individuals cannot tolerate prolonged exposures to 4-5 ppm FA. Exposure to 50-100 ppm FA for a period of 5-10 min causes serious injury to the lower respiratory tract and chronic pulmonary obstruction (IDLH, 1994).

FA has a negative influence on human health, especially on the central nervous system, the blood and the immune system. It is a potent nasal irritant, causes stunted growth, blindness and respiratory diseases. A negative impact of elevated FA content in the atmospheric air was demonstrated for children living under these conditions. A high FA level in tested blood samples correlates with low glutathione peroxidase as well as glucoso-6-phosphate dehydrogenase activities and high glutathione-S-transferase activity, thus causing the dysfunction of the glutathione system of the child's body (Zajtseva et al., 2014).

The effect of FA on human health was studied in a group of chemical industry employees who worked under conditions of frequent exposure to a combination of factors, including methanol and FA. The serum methanol and FA levels were found to be increased, and resulted in moderate multifunctional-type cognitive disorders among members of this group compared to the reference group (Maliutina and Taranenko, 2014).

Release of FA vapors in mobile homes has been associated with headache and pulmonary and dermal irritation (Ellenhorn and Barceloux, 1988). The relation between chronic respiratory symptoms and pulmonary function and FA in homes was studied in a sample of

298 children (aged 6-15) and 613 adults. The permissible FA level in industrial areas is 0.5-2.0 ppm. Wood treated with urea-formaldehydeUF resin significantly improved the mechanical properties, but FA, as well as other volatile organic compounds emitted from treated wood, have led to poor indoor air quality and a deleterious effect on human health (Wang et al., 2014).

In spite of its toxicity and negative environmental and health effects, FA is used in medicine as a preservative in the production of vaccines instead of the harmful merthiolate that can cause neurodevelopmental disorders, including autism and autism spectrum disorders (Offit and Jew, 2007; Geier and Geier, 2004, Wilton et al., 2014). FA is also applied as an analgesic agent in veterinary medicine (Schneider et al., 2014).

The development of methods for FA analysis is thus important for the control of many industrial goods, medicinal products and foodstuffs, as well as for environmental monitoring.

METHODS OF FA ASSAY

There are many chemical methods for the determination of FA (Nash, 1953; Sawicki et al., 1962; Sibirnyi et al., 2005; Bakar et al., 2009). 4-amino-3-hydrazino-5-mercapto-1,2,4-triazole (AHMT) was proposed for a FA assay (Avigad, 1983; Jung et al., 2001).

FA and other aldehydes form products of different colors, which can be selectively tested spectrophotometrically. HPLC in conjunction with steam distillation and 2,4-dinitrophenylhydrazine derivatization (2,4-DNPH) displayed good selectivity, precision and accuracy (Li et al., 2007). A polarographic method has been developed for the determination of FA traces by direct *in situ* analyte derivatization with (carboxymethyl)trimethyl ammonium chloride hydrazide (Chan and Xie, 1997).

The drawbacks of this method are that it requires expensive apparatus as well as the removal of oxygen traces by sparging with pure nitrogen. A flow injection analysis (FIA) system with an incorporated gel-filtration chromatography column was proposed and has been applied to determine FA using FdDH (Benchman, 1996).

A simple on-site rapid quantification method for trace volatile FA in aquatic products was developed by a derivative reaction-based surface-enhanced Raman spectroscopy (SERS) technique coupled with a homemade portable purge-sampling device. Trace FA separated from complicated aquatic matrices via a purge-sampling procedure was reacted with a derivative reagent to produce a Raman-active analyte for consequent SERS analysis (Zhang et al., 2014). The plasmonic properties of gold nanostructures (AuNPs) were very sensitive to the presence of FA in their surrounding medium by showing the change in the plasmonic peak's position as well as its intensity (Nengsih et al., 2012).

We have previously described several biosensoric methods (Sibirny et al., 2011a). FA-selective biosensors are based on cells (Korpan et al., 1993 and 2000; Paryzhak et al., 2008) or enzymes used as biorecognition elements: AOX (Korpan et al., 1997 and 2000; Dzyadevych et al., 2001; Alpeeva et al., 2005; Smutok et al., 2006; Shkotova et al., 2006; Shumakovich et al., 2007) or FdDH (Sibirny et al., 2011a). Potentiometric biosensors consisting of a pH-sensitive field effect transistor as a transducer and either the enzyme AOX or permeabilised yeast cells (containing AOX) as the biorecognition element have been described by Korpan et al. (2000). These biosensors have demonstrated a high selectivity to

FA with no interference response to methanol, ethanol, glucose or glycerol. Amperometric biosensors have been suggested for determination of the FA level using FdDH (Winter and Cammann, 1989; Hall et al., 1998).

Conductometric enzymatic biosensors based on FdDH (Vianello et al., 2007) and AOX (Dzyadevych et al., 2001) have been developed for FA assay.

ENZYMATIC METHODS FOR DETECTION OF FA

The cells of the thermotolerant methylotrophic yeast *Hansenula polymorpha* (*Pichia angusta, Ogatae polymorpha*), as well as FA-oxidizing enzymes isolated from these cells, were employed as the catalytic elements in the development of enzymatic methods for FA detection by analytical kits and biosensors (Sibirny et al., 2011a and b). Recombinant yeast strains originating from NCYC 495 (*leu*1-1) and CBS 4732 *H. polymorpha* (*leu*2-2) that are resistent to elevated concentrations of FA in the medium and over-produce a thermostable homologous FdDH were constructed.

The *H. polymorpha FLD*1 gene with its own promoter was inserted into the integrative plasmid pYT1 containing the *LEU*2 gene of *Saccharomyces cerevisiae* in order to construct the strains over-producing FdDH. The constructed vector was used for multi-copy integration of the target gene into the genome of *H. polymorpha* by transformation of recipient cells (Demkiv et al., 2005). Among more than 150 integrative Leu^+-transformants with a higher tolerance to FA, 14 stable clones that are resistant to 15-20 mM FA on plates were selected and studied in greater detail. Finally, FdDH activity was tested in cell-free extracts of the best FA-resistant Leu-prototrophic transformants. The stable recombinant strains Tf 11-6, originating from *leu*1-1, have the highest FdDH activity (up to 4.0 U mg^{-1}) in CE, being cultivated in a medium with 1% methanol and added FA. This value is a 4-5 fold increase compared with the parent strain. Tf 11-6 was thus characterized and chosen as a source of FdDH (Demkiv et al., 2005 and 2007). Southern dot-blot analysis revealed 6-8 copies of the *FLD1* gene in the Tf 11-6 genome, which is consistent with results obtained using the Southern-hybridization method (Demkiv et al., 2011).

A limited number of publications on the isolation and characterization of FdR and the absence of a corresponding commercial preparation of the enzyme led us to perform the screening of potential microbial FdR producers among wild-type and recombinant strains of the yeast *H. polymorpha.* The gene-engineered integrative transformants with the highest FA resistance originating from *leu* 2-2 were shown to over-produce NADH-dependent FdR upon cultivation on 1% ethanol or glycerol. The best integrative clone T 22-126 was chosen as the source for FdR isolation, and optimal cultivation conditions for the highest yield of FR were established (Demkiv et al., 2011).

The simple schemes for isolation of FdDH (Demkiv et al., 2007; Gayda et al., 2008) and FR (unpublished data) from the corresponding over-producing yeast cells and chromatographic purification of enzymes on anion-exchange sorbent were proposed, resulting in electrophoretically homogeneous enzyme preparations. AOX was isolated from the mutant strain *H. polymorpha* C-105 (*gcr1 catX*), which is defective in glucose repression of AOX synthesis and lacks catalase (Gonchar et al., 2005; Shleev et al., 2006).

The enzymatic methods and analytical kits for the FA assay were developed based on FdDH (Demkiv et al., 2007 and 2009; Gayda et al., 2008) and AOX (Gonchar et al., 2005). In methylotrophic yeasts, FdDH catalyzes the oxidation of FA to formic acid under simultaneous reduction of NAD^+ to NADH. The proposed enzymatic method is based on the photometric detection of a colored product, formazan, which is formed from nitrotetrazolium blue in a reaction coupled with FdDH-catalyzed oxidation of FA in the presence of an artificial mediator, PMS (Demkiv et al., 2007 and 2009).

We used purified preparations of FdDH and AOX, as well as yeast cells overproducing these enzymes, for construction of FA-selective electrochemical biosensors. A number of FA biosensors with different types of signal detection (amperometric, capacitance, impedance and conductometric) were developed, namely, microbial biosensors based on intact and permeabilized yeast cells (Korpan et al., 1993 and 2000; Paryzhak et al., 2008; Khlupova et al., 2006, 2007a and b2007) as well as enzyme-based sensors (Shkotova et al., 2006; Smutok et al., 2006; Nikitina, et al., 2007; Ali et al., 2006 and 2007; Dmytruk et al., 2007; Demkiv et al., 2008; Gayda et al., 2008; Sibirny et al., 2011a). All constructed biosensors were characterized by high storage and good operational stability, high sensitivity, broad dynamic range and a very low applied potential compared to known biosensors.

Comparative analysis of FA-sensitive biosensors (amperometric, capacitance, impedance and conductometric) was presented in our previous review (Sibirny et al., 2011a).

FA Analyses in Real Samples

The reliability of the proposed analytical approaches was tested on model solutions and on real samples of FA-containing industrial products, vaccines and wastes. A strong cross-correlation was found between FA content values obtained by the developed methods and by routinely used chemical methods.

The FdDH- and cells-based approaches to monitoring FA were applied for FA testing in different samples (wastewater, indoor air, industrial products, vaccines, and fish food). FdDH-based biosensors have a higher selectivity to FA than the cells-based analogues. A good correlation was observed between the data of FA testing by the FdDH- and cells-based amperometric biosenors, enzymatic and standard chemical methods (Sibirny et al., 2011a).

The FdDH- and AOX-based methods were tested on real FA-containing wastewater samples. The comparison of FA concentrations using the FdDH- and AOX-based and two routinely used chemical (chromotropic acid and MBTH) methods showed a good correlation between the four approaches. We used a standard addition test (SAT) in order to evaluate the possible interfering effect of the components of wasterwater samples on the FA assay by the FdDH-based and chromotropic acid methods.

The FdDH-based method was tested on different FA-containing vaccines (Paryzhak et al., 20072008; Sibirny et al., 2011a). As shown in Table 1, a comparison of the FA concentration obtained by the enzymatic method and by two routinely used chemical ones showed a good correlation between the two approaches.

Lower levels of FA in vaccines ADP-M and AD, obtained using the enzymatic method as compared to the chemical methods, may be due to the inhibitory effect of the mercury-containing vaccine preservative 0.01% merthiolate on the enzyme. The AOX-based method exploits an aldehyde-selective reagent, 3-methyl-2-benzothiazolinehydrazone (MBTH) (Sibirny et al., 2008). This enzymo-chemical method of differential detection of FA and methanol in mixtures was used to analyze samples of a commercial product, formalin and samples of wastes. The results of this analysis were in good agreement with data obtained by traditional chemical methods and gas chromatography (Sibirny et al., 2011b, 2011a). The proposed method, contrary to the standard chemical approach, does not require the use of aggressive chemicals (concentrated sulfuric, phosphoric and chromotropic acids, or permanganate), it is easier to perform, and can be used for industrial waste verification and certification of formaline-containing materials.

AOX- and FdDH-based methods were used for FA assay in fish food products (Sibirny et al., 2011a2011b). A good correlation was found between the analytical results of both enzymatic methods compared with chemical approaches, although the AOX-based assay is preferrable due to its higher sensitivity, good linearity, insensitivity to interference by test sample contaminants and usage of non-aggressive reagents for the sample's pre-treatment and assay procedure (Tables 1 and 2).

A laboratory prototype of the FA analyzer with a catalytic chemosensing element and a simple low-cost FA-sensitive biosensor for semi-quantitative FA detection were developed. The latter sensor is based on a change in the color of a solution that contains a mixture of AOX, horseradish peroxidase and a chromogen after exposure to airborne FA. Selectivity for the target analyte could be increased by modulating the working electrode potential. Analytical parameters of three variants of the amperometric analyzer that differed in the chemical structure/configuration of the working electrode were studied.

Table 1. Comparison of FA assay methods (mg/L) in vaccines

Vaccine/Method	Enzymatic	Chemical	
	FdDH-based	Chromotropic acid	MBTH
Anti-diphtheria	15±2.5	36±1.4	31±2.0
Anti-diphtheria and tetanus	17±1.9	27±2.0	29±2.7
Polio-vaccine "Imovax"	30±3.0	27±2.6	–
Tetanus	10.2±0.6	9.0±0.2	12.0±0.2

Sibirny et al., 2011a.

Table 2. Results of FA assay (in mg per 1 kg of wet weight of muscle tissue) in protein-free extracts of fish using three independent approaches: AOX-based method, Nash's method and Purpald's method

Fish / Approach	Hake	Cod	Carp
AOX-based	90.0±2.6	74.1±1.1	0
Nash's	121.6±0.9	96.8±3.1	0
Purpald	100.6±1.6	59.7±3.1	0

The constructed analyzer was tested on FA-containing wastewater solutions (Sigawi et al., 2014a). The chemosensing electrode for FA analyses was modified with platinum black as a catalyst according to (Kovalyshyn et al., 2010).

Optimization of the working potential for chemoselective oxidation of FA also included a study of selectivity.

During the preliminary stages of the research, it was found that the sensor exhibited a higher response (selectivity) for FA at a working potential of +50 mV versus Ag/AgCl, and a lower response for methanol and phenol. A significant increase in the sensor's selectivity was achieved at lower potentials of the working electrode (Figure 1).

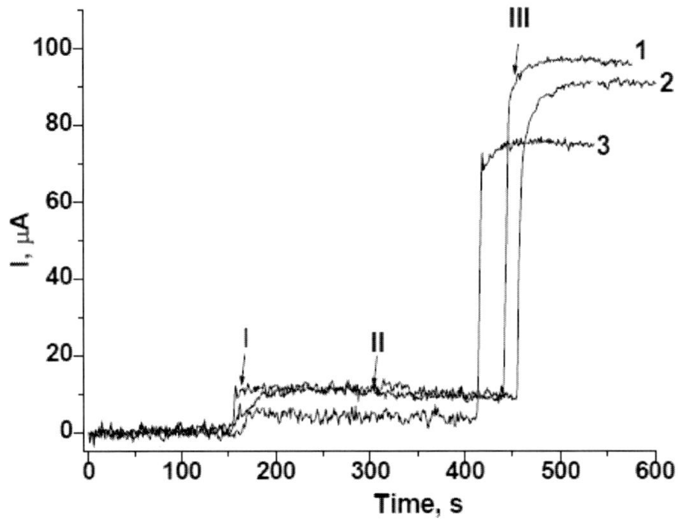

a

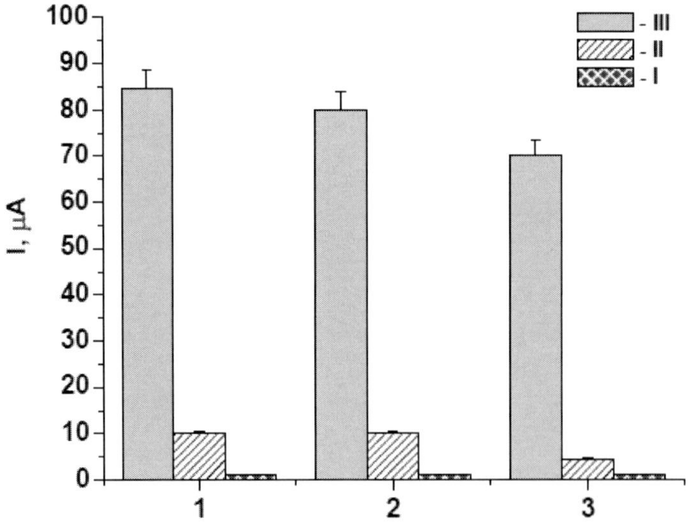

b

Figure 1. Chronoamperometric (Aa) and amperometric response (Bb) of the chemosensor based on a 3.05 mm rod platinized carbon electrode registered in 25 mM phosphate buffer, pH 8.0 under different working potentials: +50 mV (1), 0 mV (2) and -50 mV (3) to various 2.5 mM analytes: methanol (I), phenol (II) and FA (III).

The proposed chemosensor has some advantages when compared with biosensors, namely, high stability, simplicity of electrode preparation and ability to carry out the analyses under low potentials. The last property is important for analysis of complex real samples.

FA analyses using a chemosensor based on a platinized carbon electrode (3.05 mm-diameter rod) were carried out in real wastewater samples (Figure 2), a formalin solution (Figure 3) and hake fish (Figure 4).

In all cases, the FA analyses were carried out using a variant of standard addition tests. The calibration curve was obtained by adding aliquots of a FA standard solution to the diluted analyzed sample.

Parameters of the line ($Y=A+B\cdot X$) are shown in the inserts: "A" - current, corresponding to the signal of the analyzed sample without addition of exogenous FA standard, "B" - slope of the calibration curve, and "R" - linear regression coefficient.

Taking the values of the parameters of the linear regression (A, B) and the dilution factor of the tested sample (n) into account, the initial concentration of the tested sample (C_o) may be calculated according to the equation: $C_{sample}=A/B\times n$.

The results of quantitative FA determination by several methods in real samples are summarized in Tables 1-4. The reproducibility of the proposed chemosensoric method is satisfactory. A strong correlation ($R_{ave}=0.99$) was shown to exist between the values obtained by the elaborated chemosensor and other analytical methods (Figure 5).

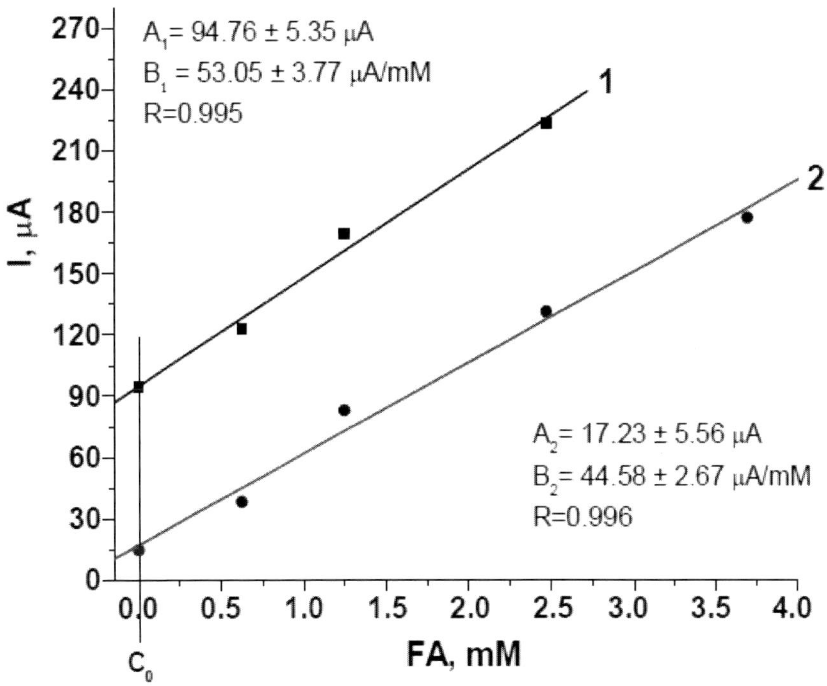

$n_1=200$; $C_1=360$ mM; $n_2=1000$; $C_2=386$ mM.

Figure 2. Calibration curves for FA analysis of the wastewater sample with the chemosensor. The dilution factors (n), the calculated initial concentration (C_o) and statistical data such as parameters of linear regression (coefficients of the equation $Y=A+B\cdot X$, where Y– current A; X– FA concentration, mM; A– current corresponding to the variant without addition of exogenous FA; B– slope value; and R– linear regression coefficient) are indicated. 1 and 2 – different dilutions of the tested sample. n1=200; C1=360 mM; n2=1000; C2=386 mM.

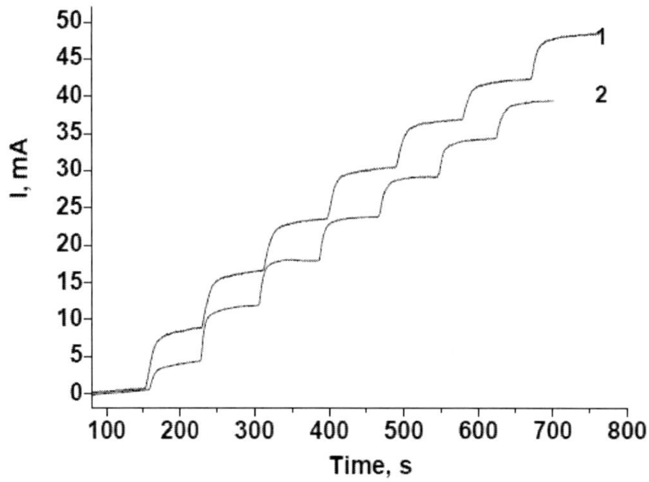

a

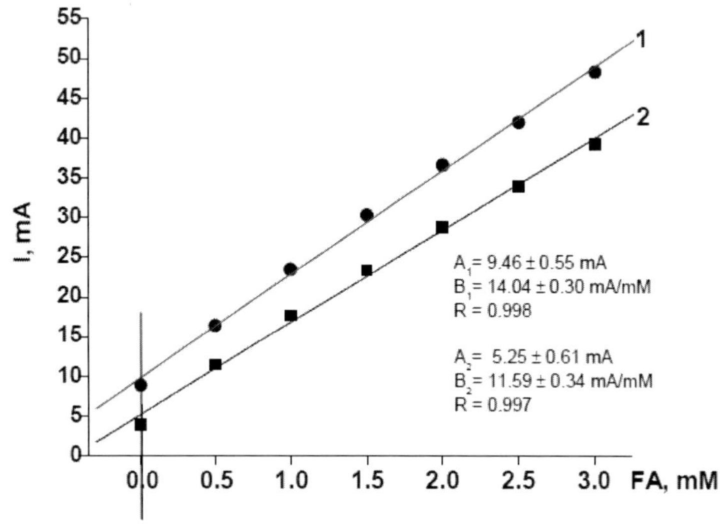

b

Figure 3. Chronoamperogramm (Aa) and calibration graphs on FA (Bb) of the chemosensor for formalin analyses registered in 25 mM phosphate buffer, pH 8.0. 1 and 2 – different dilutions of the tested sample. n_1=2000, C_1=13.51 M; n_2=3000, C_2=13.59 M.

METHODS OF FA REMOVAL

FA is regarded as the major indoor pollutant emitted from widely used building and decorative materials in airtight buildings, which should be eliminated under indoor environmental conditions. Many novel approaches for detoxification of FA have been developed to date based on the usage of various filters and adsorbents (Cai and Wu, 2014; Coburn, 2014; Duan, L., 2014; Guo, 2014; Hong and Wang, 2014; Tian and Gao, 2014; Liu S., 2014; Liu and Yang, 2014; Liu Z., 2014).

The design and fabrication of high performance adsorbents for removing indoor FA based on TiO_2 nanosheets (Zhou et al., 2013) as well as a Pt/TiO_2 composition (Nie et al.,

2013) were proposed. The functionalized meso-silica materials and new green nanosilica materials were studied and tested as adsorbents for FA vapor from contaminated air (Ewlad-Ahmedet al., 2012).

A combined reaction consisting of granular activated carbon (GAC), adsorption and catalytic oxidation has been proposed for improving FA removal.

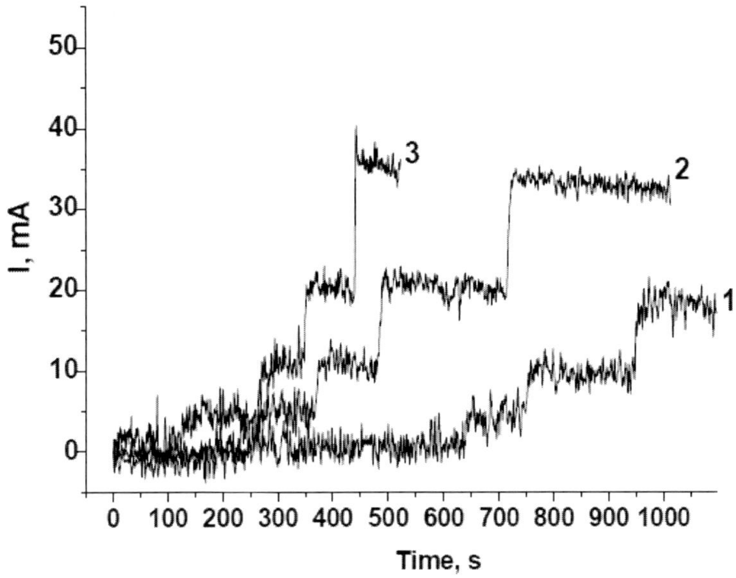

a

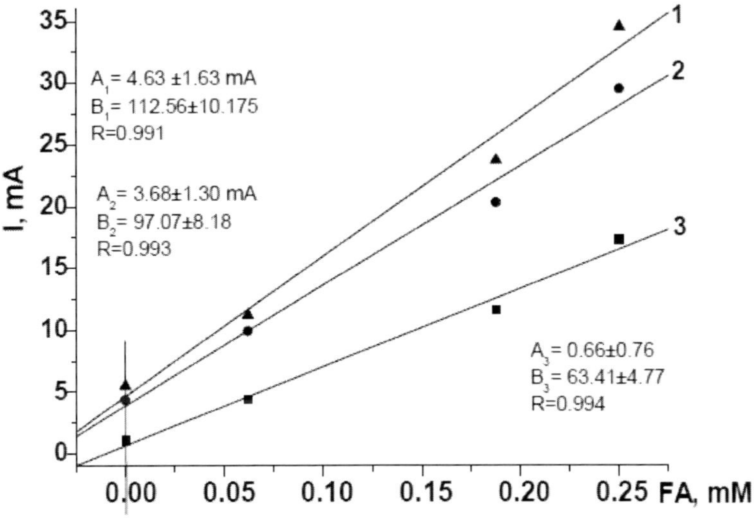

b

Figure 4. Chronoamperogram (Aa) and calibration graphs on FA (Bb) of the chemosensor registered in 25 mM phosphate buffer, pH 8.0 in analyses of hake fish under different dilutions: n_1=1,7, n_2=2 and n_3=6. Calculated concentrations in the samples: C_1=0,069 mM; C_2=0,075 mM; C_3=0,060 mM.

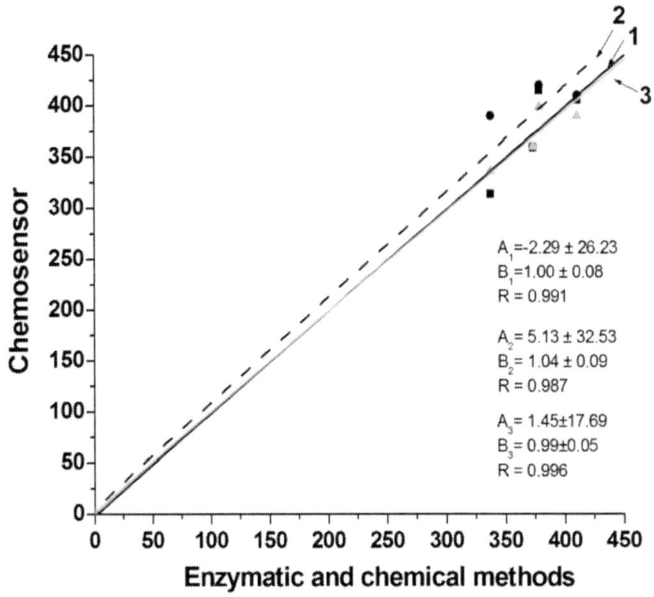

Figure 5. Correlations between the results of FA estimation in wastewater by different methods: 1 – enzymatic FdDH-based method, 2 – chemical method with chromottropic acid, and 3 – chemical method with MBTH, relative to the proposed chemosensor's data.

Table 3. FA concentration (mM) in a sample of formalin estimated with different methods

FA analyses	Methods						
	Chemical				Enzymatic		
	Chromotropic acid	MBTH	Purpald	Chemosensor	FdDH-based biosensor		Cell biosensor
					automatic	manual[1]	
Concentration, mM	14±2.4	12.6±2.1	12.9±1.9	13.6±1.8	13.5±2.1	13.6±0.6	13.8±0.5
Reference	Demkiv et al., 2008			(this paper)	Demkiv et al., 2008		

[1]biosensor, integrated in automatic analyzer "OLGA".

Table 4. FA concentration (mM) in a sample of wastewater determined with different methods

FA analyses	Methods			
	Chemical			Enzymatic
Standards added in the tested sample	Chemosensor	Chromotropic acid	MBTH	FdDH
FA	373 ± 19	360 ± 6	360 ± 7	359 ± 9
FA and methanol	337 ± 19	390 ± 12	336 ± 17	314 ±12
FA and phenol	378 ± 4	420 ±8	399 ± 16	415 ± 17
FA, methanol and phenol	410 ± 10	410 ± 11	390 ±16	406 ±17

Silver nano-particles (AgNPs) attached onto the surface of GAC (Ag-GAC) using the sputtering method were evaluated for the simultaneous catalytic oxidation and adsorption of FA. The appearance of CO_2 from the AgNPs indicated that FA was converted to its final product, with the oxidation kinetics expressed as pseudo-first order (Shin and Song, 2011).

A catalytic oxidation process of FA over mesoporous Co_3O_4, Co_3O_4-CeO_2, Au/Co_3O_4, and Au/Co_3O_4-CeO_2 catalysts and their excellent catalytic performances at room temperature was investigated. These catalysts were prepared by a "nanocasting" method with the mesostructure being generated from SBA-15 silica with a 2D structure (Ma et al., 2011).

Control of the FA concentration requires the development of simple, cheap, reliable, sensitive and selective methods for FA assay. The development of such methods for fast and inexpensive FA detection is very important for analytical biotechnology, including environmental monitoring and bioremediation. To solve this problem, we developed analytical approaches for FA assay based on the yeast enzymes NAD^+- and glutathione-dependent FdDH (hereinafter: FdDH) as well as on alcohol oxidase (AOX) isolated from overproducing yeast cells.

BIOREMEDIATION OF FA

The main approaches to bioremediation of toxic wastes are described in a recent review by Kumar et al., (2011), where the term bioremediation was defined as a process of toxic waste removal from the environment using biological agents. Bioremediation, which can be performed *in situ* and *ex situ*, is considered an attractive approach for cleaning polluted environments, including wastewater and recovery of contaminated soil.

PLANTS FOR BIOREMEDIATION OF FA

The efficiency of FA removal was tested in 86 species of plants representing five general classes, namely, ferns, woody foliage plants, herbaceous foliage plants, Korean native plants and herbs (Kim et al., 2010). Phytoremediation potential was assessed by measuring the rate of FA removal by the plants exposed to gaseous FA (2.0 $\mu L \cdot L^{-1}$) in airtight chambers (1.0 m^3) constructed of inert materials. The following species: *Osmunda japonica, Selaginella tamariscina, Davallia mariesii, Polypodium formosanum, Psidium guajava, Lavandula spp., Pteris dispar, Pteris multifida, and Pelargonium spp.* were found to be the most effective, removing more than 1.87 $\mu g(FA) \cdot m^{-3} \cdot cm^{-2}$ over 5 h.

Transgenic plants able to oxidize higher levels of aqueous and gaseous FA to formate than the wild-type plants may be a prospective for improving the phytoremediation of FA pollution. The *faldh* gene, which codes for a putative *Brevibacillus brevis* FdDH, was isolated and then transformed into tobacco.

Overexpression of *faldh* in transgenic *tobacco* (Nian et al., 2013) and *Arabidopsis* (Achkor et al., 2003) confers a high tolerance to exogenous FA and an increased ability to take up FA. The capacity to take up and detoxify high concentrations of FA is proportionally related to the FdDH activity in the plant, revealing the essential role of this enzyme in FA detoxification by "green" approaches.

An environmentally friendly "green" and pollution-free method of FA removal from plates was based on the use of plant extracts and NPs of titanium oxide. The mixed plant extract consists of aloes, chlorophytum comosum and cactus. Spraying the extract mixture with NPs agent on the surfaces of the plates followed by ultraviolet irradiation resulted in FA removal from plates (Yang, 2014).

MICROBIOGICAL BIODDEGRADATION OF FA

Biodegradation of FA can be performed with the help of methylotropic and autothropic microorganisms (bacteria, yeast, fungi and cyanobacteria) using FA as the sole carbon and energy source. The list of microorganisms capable of FA degradation includes *Pseudomonas spp.* generum (Kato et al., 1983), *Halomonas spp.* (Azachi et al., 1995) various strains *of Methylotropha* (Attwood and Quayle, 1984), *Debariomyces spp.* and *Trichosporon spp.* yeast genera (Kato et al., 1982), *Hansenula spp.* (van Dijken et al., 1975, 1976), *Candida spp.* (Pilat and Prokop, 1976) and *Gliocladium spp.* fungi (Sakagushi et al., 1975).

The biodegradation mechanisms are based on anabolic or catabolic involvement of FA in cell processes (van Agteren et al., 1998). The catabolic pathway assumes enzymatic oxidation of FA into formic acid and then to carbon dioxide by enzymes such as FdDH, aldehyde dehydrogenase, formaldehyde dismutase, methylformate synthase and AOX (van Agteren et al., 1998).

Eiroa et al. studied simultaneous removal of FA and ammonium in activated sludge cultures when FA alone or FA mixed with methanol served as the sole carbon source. In the first case, the authors report initial consumption rates between 0.22 and 8.2 g (FA) g^{-1} (VSS, volatile suspended solids) d^{-1} when the FA concentration ranged from 30 to 3890 mg dm^{3}. In the presence of 12.5% methanol, the FA biodegradation rates were even higher: the degradation rate for the 2000 mg (FA) dm^{-3} concentration was 4.6 g (FA) g^{-1} (VSS) d^{-1} and upon addition of methanol it increased to 7.3 g (FA) g^{-1} (VSS) d^{-1}. FA was found to inhibit a nitrification process when complete inhibition was observed at FA concentrations above 1500 mg dm^{-3}.

The authors also performed continuous simultaneous degradation of FA and ammonium by anoxic sludge from a full-scale plant that treats wastewater of a factory producing synthetic resin. The FA biodegradation rate in the batch reactor reached 3.3 g (FA) g^{-1} $(VSS)^{-1}$ d^{-1} at a FA concentration of 2150 mg L^{-1} and in the continuous reactor, the FA removal efficiency was almost 99% at FA loading rates between 625-5000 mg (FA) L^{-1}. The FA degradation products were methanol and formic acid. The studies of Eiroa et al. were targeted to solving the problem of remediation of wastewater from synthetic resin-producing industries that contains high concentrations of both FA and ammonia (Eiroa et al., 2004 and 2007).

Omil et al. (1999) described biodegradation of FA in anaerobic batch digesters by un-adapted sludges from an industrial wastewater plant for treatment of effluencts from a seafood processing factory in the presence and absence of volatile fatty acids, added as co-substrates. Biodegradation of FA was enhanced in the presence of co-substrates, especially acetate. Methanol was the main intermediate of anaerobic degradation of FA in all cases, with or without co-substrates.

Biodegradation of FA by municipal wastewater bacteria was studied by Lotfy and Rashed (2002). The authors report temporary or almost complete inhibition of FA biodegradation depending on the FA concentration. Addition of sodium sulphite to the FA-containing effluents in a 1:1 FA to sulphite mole ratio prevented inactivation of bacteria and enabled the achievement of high FA degradation rates.

Mitsui's (Mitsui et al., 2006) group studied FA consumption by two methylotrophic bacteria – *Methylobacterium* sp. MF1 (an obligatory methylotroph) and *Acidomonas methanolica* MB58 (a facultative methylotroph), which showed similar efficiencies of FA biodegradation (1.2 g $L^{-1}d^{-1}$ and ca. 1.8 g $L^{-1}d^{-1}$, respectively) under continuous cultivation in a chemostat. By measuring key enzyme activities in each examined strain, the authors found that FA was assimilated by different pathways: in the former strain the consumption occurred via the serine pathway, whereas in the latter it occurred via the ribulose-monophosphate pathway.

Biodegradation of FA by *Pseudomonas putida* was studied by Adroer et al. (1990). It was shown that degradation occurred by a dismutation pathway known as Cannizaro's reaction, yielding methanol and formic acid as products catalyzed by the enzyme formaldehyde dismutase. The authors showed that biodegradation of FA can be performed in a fluidized bed bioreactor and used for treatment of industrial wastewater.

Kaszycki and Koloczek (2002) showed that the methylotrophic yeast *Hansenula polymorpha* can cooperate with activated sludge taken from wastewater treatment stations and enhance biodegradation of FA in industrial wastewater. The authors propose treating FA-containing wastewater using biological filters that contain separate chambers with methylotrophic yeast monocultures or by introducing methylotrophic yeast into activated sludge in aeration chambers of existing biological wastewater treatment stations maintained at different chemical plants.

Several FA-resistant yeasts, identified as strains of *Debaryomyces vanriji* and *Trichosporon penicillatum*, were isolated from soil by Kato et al. (1982). The cells were shown to consume FA when its concentration in the growth medium was in the range of 0.15 to 0.55%. The authors related the yeasts' FA resistance to their FA-degrading activity associated with high enzymatic activity in cell-free extracts containing FA dehydrogenase, S-formylglutathione hydrolase and formate dehydrogenase. The presence of these enzymes indicated that FA underwent sequential oxidation to CO_2.

Selected strains of *Pseudomonas putida*, *Pseudomonas cepacia, Trichosporon penicillatum* yeast and a mixed culture of these three species were used for aerobic degradation of FA and formic acid in synthetic medium and wastewater generated by melamine resin production (Glancer-Šoljan et al., 2001). Microorganisms such as *Methanosaeta* probably survived due to their preferential use of acetate, while *Methanosarcina* preferentially uses methanol (Oliveira et al., 2004). Biodegradation of FA was also tested in the marine microalga *Nannochloropsis oculata* (Yoshida et al., 2009).

Microorganisms utilizing methanol have adopted several metabolic strategies to overcome FA toxicity. Mechanisms of FA detoxification in yeast, bacteria and archaea were studied (Yurimoto et al., 2005). The toxicity of FA in batch assays, using volatile fatty acids as co-substrates and the continuous anaerobic treatment of FA-containing wastewaters in upflow anaerobic sludge blanket reactors, was investigated (Vidal et al., 1999). The kinetic process of FA biodegradation in a biofilter packed with a mixture of compost, vermiculite powder and ceramic particles was investigated by Xu et al. (2010).

Enzymatic degradation of FA by AOX purified from the formaldehyde-resistant fungus *Paecilomyces variotii* IRI017 isolated from FA-containing wastewater is described by Kondo et al. (2008). AOX was found to oxidize not only methanol and ethanol, but also lower primary aliphatic alcohols and FA. AOX's ability to oxidize FA to formic acid was enhanced upon addition of catalase which eliminated hydrogen peroxide. The catalytic characteristic of the isolated AOX resembled the activity of AOX from methylotrophic yeasts.

Another FA-oxidizing enzyme was isolated from the membrane of *Acetobacter* sp. SKU 14 (Shinagawa et al., 2006). The enzyme was found to be NAD(P)-independent and was identified as a quinoprotein alcohol dehydrogenase (QADH). The QADH exhibited specificity towards a broad spectrum of substrates: alcohols, including n-butanol, n-propanol and ethanol, and aldehydes. It should be mentioned that this was the first report of the ability of QADH to effectively oxidize FA. The authors attribute FA oxidation with a central role in the metabolism of methanotrophic and methylotrophic bacteria, since FA is an intermediate in metabolic chains of the one-carbon compounds methane, methanol and methylamine.

Our group studied biodegradation of airborne and waterborne FA by the methylotrophic yeast *H. polymorpha* and by the enzymes AOX and FdDH isolated from yeast cells. It was shown that airborne FA can be oxidized by AOX isolated from a *H. polymorpha* C-105 mutant (*gcr1 catX)* overproducing this enzyme in glucose medium (Sigawi et al., 2011; Nisnevitch et al., 2011) The isolated enzyme was immobilized in calcium alginate gel and the gel beads were applied into a continuous fluidized bed bioreactor. Air containing 0.3 to 18.5 ppm FA was bubbled through immobilized AOX and the concentration of FA in the outlet gas phase decreased to 0.02-0.03 ppm, i.e., the FA concentration was ten-fold lower than its threshold limit value (Sigawi et al., 2011; Sibirny et al., 2011a). Similar results were obtained with another enzyme, FdDH, a NAD- and glutathione-dependent enzyme isolated from a recombinant Tf 11-6 strain of the methylotrophic yeast *H. polymorpha.* The bioreactor with immobilized FdDH provided 97-100% elimination of airborne FA (Sigawi et al., 2012). The *H. polymorpha* cells also exhibited a high ability to decompose FA. Permeabilized *H. polymorpha* C-105 cells and intact cells of the Tf 11-6 strain were immobilized in calcium alginate and tested for continuous oxidation of airborne FA. It was shown that the former cells provided ca. 90% elimination of airborne FA and the latter decomposed 94-100% of the FA (Sigawi et al., 2011 and 2012). Immobilized AOX was combined with immobilized aminoxidase (AMO), a copper-containing enzyme from the thermotolerant yeast *H. polymorpha*, and a two-step scheme for bioconversion of airborne methylamine and generated FA was developed (Sigawi et al., 2014b). Airflow containing methylamine was fed into a fluidized bed bioreactor with immobilized AMO and the outlet of the first reactor was connected to the inlet of the second reactor which contained immobilized AOX. Methylamine biodegradation resulted in the formation of FA, which was in turn degraded in the second bioreactor. The proposed system provided almost total elimination of the airborne methylamine and the FA. We used a multi-chamber airlift reactor supplied with a two-channel peristaltic pump for continuous feed and discharge of an aqueous phase for continuous biodegradation of waterborne FA (Figure 6). Each second chamber of the reactor was connected to a source of air for effective but non-violent mixing. Cartridges made from a polymer mesh-work having 1 mm pores were installed into each chamber not connected to the airstream, and beads of calcium alginate gel with the immobilized AOX were placed inside the cartridges. As in the previous studies, AOX was isolated from *H. polymorpha* C-105 cells.

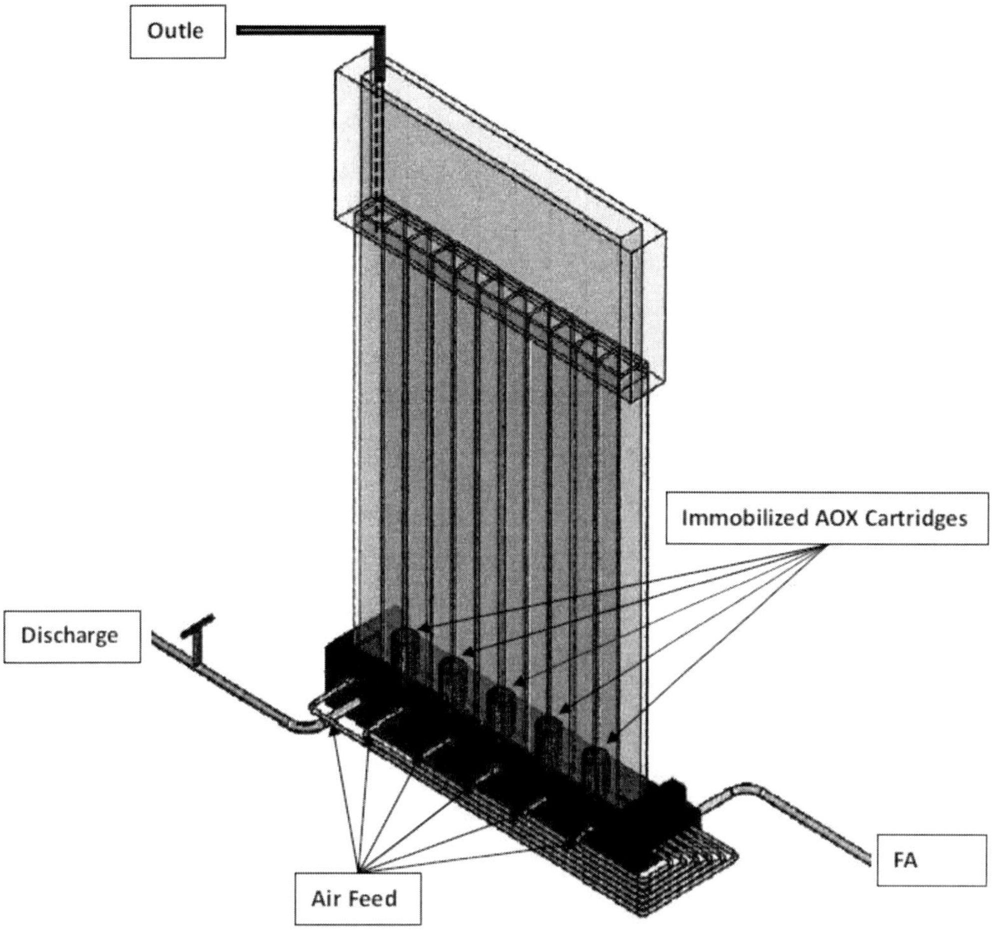

Figure 6. Multi-chamber airlift bioreactor for continuous oxidation of FA by AOX immobilized in calcium alginate gel. The reactor had dimensions of H948xW550xD29 mm at operating volume of 12.7 L, air pressure was 20 psi.

Waterborne FA had free access to the immobilized AOX and the products of the reactions could leave the gel without hindrance due to transparency of the cartridges for the aqueous phase and continuous mixing. The FA concentration was monitored at the inlet and outlet of the reactor. In control experiments, waterborne FA was flowed through the reactor at the same concentrations and flow rates as in the experiments on biodegradation. Figure 7 shows results of the FA degradation by immobilized AOX when the inlet FA concentration was 1.1 mM. In the control experiments, the FA outlet concentration was the same as at the inlet, but when FA solutions were passed through the reactor with immobilized AOX, the outlet FA concentration decreased in time up to the zero value. The time needed for reaching an equilibrium condition depended on the flow rates, when the shortest equilibrium time was 9 h at the lowest flow of 7 mL min^{-1} and the longest time was 25 h for a flow of 18 mL min^{-1}. FA was not found in the outlet flow once equilibrium was reached (Figure 7).

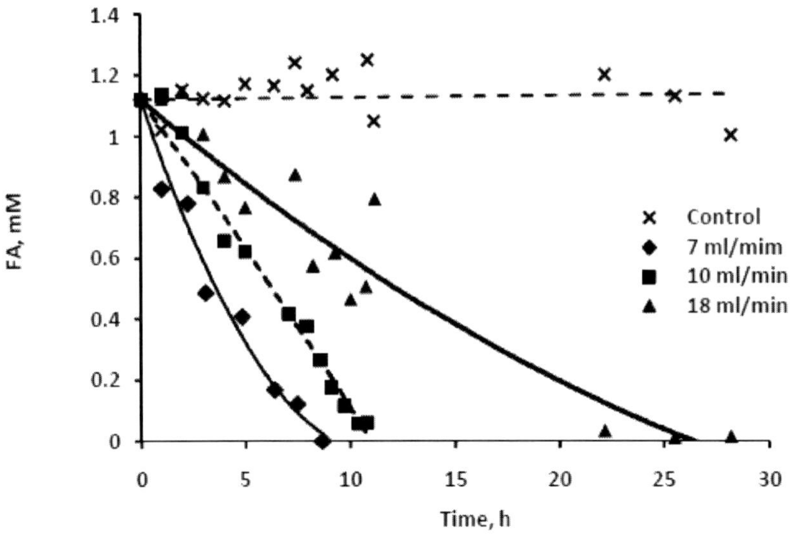

Figure 7. Continuous degradation of waterborne FA by AOX immobilized in calcium alginate gel. The immobilized AOX (6.6 U per g of the gel) was divided into 16.6 g portions into 5 polymeric networks having 1 mm pores. 1.1 mM FA solution was fed into the airlift bioreactor at flows of 7, 10 or 18 ml min^{-1}. The FA concentration was monitored at the inlet and outlet from the reactor as shown in Figure 6.

CONCLUSION

Biological approaches to detection and degradation of FA can be considered very promising.

To improve the existing technologies the following aspects are to be developed: extension of storage stability of enzymes, such as AOX and FdDH, as well as of recombinant microbial cells; prevention of co-enzymes' leakage from bioactive layers of biosensors; cheap fabrication of bioelectrodes with immobilized enzyme or microbial cells and design of industrial bioreactors for the FA bioremediation.

ACKNOWLEDGMENTS

This work was financially supported by NAS of Ukraine (Project 6/1-2014, program "Sensors for Medical, Environmental, Industrial, and Technological Needs") (to G.G., O.D., H.K. and M.G.), by the Cherna Moskowitz Foundation, California, US (to R.H.-L.) and by the Research Authority of the Ariel University, Israel (to D.W. and M.N.). We acknowledge Mr. Ran Klein (Supergum Company, Barkan, Israel) for drawing the air-lift reactor chart.

REFERENCES

Achkor, H., Díaz, M., Fernández, M. R., Biosca, J. A., Parés, X., Martínez, M. C., (2003). Enhanced formaldehyde detoxification by overexpression of glutathione-dependent formaldehyde dehydrogenase from arabidopsis. *Plant. Physiology.* 132, 2248-55.

Adroer, N., Casas, C., de Mas, C., Solfi,, C. (1990). Mechanism of formaldehyde biodegradation by *Pseudomonas putida. Appl. Microbiol. Biotechnol.* 33, 217-20.

Ali, M. B., Gonchar, M., Gayda, G., Paryzhak, S., Maaref, M. A., Jaffrezic-Renault, N., Korpan, Y., (2007). Formaldehyde-sensitive sensor based on recombinant formaldehyde dehydrogenase using capacitance versus voltage measurements. *Biosens. Bioelectron.* 22 (12), 2790-5.

Ali, M. B., Korpan, Y., Gonchar, M., El'skaya, A., Maaref, M. A., Jaffrezic-Renault, N., Martelet, C., (2006). Formaldehyde assay by capacitance versus voltage and impedance measurements using be-layer bio-recognition membrane. *Biosens. Bioelectron.* 22, 575-81.

Alpeeva, I. S., Vilkanauskyte, A., Ngounou, B., Csöregi, E., Sakharov, I. Yu., Gonchar, M., Schuhmann, W., (2005). Bi-enzyme alcohol biosensors based on genetically engineered alcohol oxidase and different peroxidases. *Microchim. Acta.* 152, 21-7.

Attwood, M. M., Quayle, J. R., (1984). Formaldehyde as a central intermediary metabolite of methylotrophic metabolism, In: *Microbial Growth on C1 Compounds*, R. L. Crawford, R. S. Hanson (Ed.), pp. 315-323, American Society for Microbiology, D.C., Washington.

Avigad, G., (1983). A simple spectrometric determination of formaldehyde and other aldehydes: application to periodate oxidized glycerol systems. *Anal. Biochem.* 134 (2), 499-504.

Azachi, M., Henis, Y., Oren, A., (1995). Transformation offormaldehyde by a Halomonas sp. *Can. J. Microbiol.* 41 (6), 548-53.

Bakar, A. A., Salleh, A. B., Abdulamir, Y. H., (2009). Areview: Methods of determination of health-endangering formaldehyde in diet. *Res. J. Pharmacol.* 3 (2), 31-47.

Benchman, I. E., (1996). Determination of formaldehyde in frozen fish with formaldehyde dehydrogenase using flow Injection system with an Incorporated gel-filtration chromatography column. *Anal. Chim. Acta.* 320 (2-3), 155-164.

Cai, X., Wu, T., (2014-06-18). Functional synthetic fiber capable of removing formaldehyde and preparation and application of fiber. CN103861624 (A).

Coburn, C. E., (2014-03-27). Composition and method for removing excess formaldehyde. US2014083953 (A1).

Conaway, C. C., Whysner, J., Verna, L. K., (1996). Formaldehyde mechanistic data and risk assessment: endogenous protection from DNA adduct formation. *Pharmacol. Ther.* 71 (1-2), 29-55.

Chan, W. H., Xie, T. Y., (1997). Adsorption voltammetric determination of $\mu g \cdot L^{-1}$ levels formaldehyde via in situ derivatization with Girard's reagent T. *Anal. Chim. Acta.* 339 (1), 173-9.

Dalbey, W. E., (1982). Formaldehyde and tumors in hamster respiratory tract. *Toxicology.* 24, 9-14.

Demkiv, O. M., Gayda, G. Z., Gonchar, M. V., (2009). An enzymatic method for formaldehyde assay based on formaldehyde dehydrogenase from the recombinant yeast *Hansenula polymorpha. Ukr. biochem. J.* (in Ukrainian). 81 (6), 111-20.

Demkiv, O. M., Paryzhak, S. Y., Ishchuk, O. P., Gayda, G. Z., Gonchar, M. V., (2011). Activities of the enzymes of formaldehyde catabolism in recombinant strains of *Hansenula polymorpha. // Mikrobiologiya* (Moscow). 80 (3), 307-13.

Demkiv, O. M., Paryzhak, S. Ya., Gayda, G. Z., Sibirny, V. A., Gonchar, M. V., (2007). Formaldehyde dehydrogenase from the recombinant yeast *Hansenula polymorpha*: isolation and bioanalytic application. *FEMS Yeast Research.* 7, 1153-9.

Demkiv, O. M., Paryzhak, S. Ya., Krasovs'ka, E. S., Stasyk, O. V., Gayda, G. Z., Sibirny, A. A., Gonchar, M. V., (2005). Construction of methylotrophic yeast *Hansenula polymorpha* strains overproducing formaldehyde dehydrogenase. *Biopolymers and cell.* 21(6), 525-30 (in Ukrainian).

Demkiv, O., Smutok, O., Paryzhak, S., Gayda, G., Sultanov, Y., Guschin, D., Shkil, H., Schuhmann, W., Gonchar, M., (2008). Reagentless amperometric formaldehyde-selective biosensors based on the recombinant yeast formaldehyde dehydrogenase. *Talanta.* 76 (4), 837-46.

Dijken, van, J. P., Oostra-Demkes, G. J., Otto, R., Harder, W., (1976). S-Formylglutathione: the substrate for formate dehydrogenase in methanol-utilizing yeasts. *Arch. Microbiol.* 111 (1-2), 77-83.

Dingle, P., (1997). Sick building syndrome. Advances in Environmental Control Technology: Health and Toxicology. *Advances in Environmental Control Technology Series.* 67-92.

Dmytruk, K. V., Smutok, O. V., Ryabova, O. B., Gayda, G. Z., Sibirny, V. A., Schuhmann, W., Gonchar, M. V., Sibirny, A. A., (2007). Isolation and characterization of mutated alcohol oxidases from the yeast *Hansenula polymorpha* with decreased affinity toward substrates and their use as selective elements of an amperometric biosensor. *BMC Biotechnology.* 7 (7), 1-7.

Duan, L., (2014-06-25). Method for removing formaldehyde in air. CN103877826 (A).

Dzyadevych, S. V., Arkyhypova, V. N., Korpan, Y. I., El'skaya, A. V., Soldatkin, A. P., Jaffrezic-Renault, N., Martelet, C., (2001). Conductometric formaldehyde sensitive biosensor with specifically adapted analytical characteristics. *Anal. Chim. Acta.* 445 (1), 47-55.

Eiroa, M., Kennes, C., Veiga, M. C., (2004). Formaldehyde biodegradation and its inhibitory effect on nitrification. *J. Chem. Technol. Biotechnol.* 79, 499-504.

Eiroa, M., Vilar, A., Kennes, C., Veiga, M. C., (2007). Formaldehyde biodegradation and its effect on the denitrification process. *Environ. Technol.* 28(9), 1027-33.

Ellenhorn, M. J., Barceloux, D. G., (1988). *Medical Toxicology - Diagnosis and Treatment of Human Poisoning.* New York, NY: Elsevier Science Publishing Co., Inc., p. 1002.

Ewlad-Ahmed, A. M., Morris, M. A., Patwardhan, S. V., Gibson, L.T., (2012). Removal of formaldehyde from air using functionalized silica supports. Environ. Sci. Technol. 46(24), 13354-60.

Feron, V. J., Til, H. P., de Vrijer, F., Woutersen, R. A. Cassee, F. R., van Bladeren, P. J. (1991). Aldehydes: occurrence, carcinogenic potential, mechanism of action and risk assessment. *Mutat. Res.* 259, 363-85.

Gayda, G., Demkiv, O., Paryzhak, S., Gonchar, M. (2008). Recombinant formaldehyde dehydrogenase and gene-engineered methylotrophic yeasts as bioanalitycal instruments for assay of toxic formaldehyde. In: *Algal Toxins: Nature, Occurrence, Effect and Detection*. Evangelista, V., et al. (Ed.). pp. 311-333. Series: NATO Science for Peace and Security. VIII, ISBN: 978-1-4020-8478-2.

Geier, D. A., Geier, M. R., (2004). Neurodevelopmental disorders following thimerosal-containing childhood immunizations: a follow-up analysis. *Int. J. Toxicol.* 23, 369-76.

Gerberich, H., Seaman, G., (20042007). Formaldehyde. In: *Encyclopedia of Chemical Technology*. 4th 5th edit. New York: John Wiley and Sons. Vol. 1112, 929-51107-28.

Glancer-Šoljan, M., Šoljan, V., Dragičević, T. L., (2001). Aerobic degradation of formaldehyde in wastewater from the production of melamine resins. *Food Technol. Biotechnol.* 39 (3), 197-202.

Gonchar, M. V., Grabek, D., Oklejewich, B., Pavlishko, H. M., Shamlian, O. V., Sybirny, V. A., Kotylak, Z., Rudke, K., Rudke, K., Csoregi, E., Sibirny, A. A., (2005). A new enzymo-chemical method for simultaneous assay of methanol and formaldehyde. *Ukr. Biochem. J.* 77 (3), 146-54.

Guo, J., (2014-08-06). Photocatalytic glass ball for removing formaldehyde. CN103962128 (A).

Hall, E. A. H., Preuss, M., Gooding, J. J., (1998). Exploring Sensors to Monitor Some Environmental Discharges. In: *Biosensors for Direct Monitoring of Environmental Pollutants in Field*, Nikolelis, D. P., Krull, U. J., Wang, J. and Mascini, M. (Ed.), Kluwer Academic, London. 227-237.

Hansen, J., Olsen, J. H., (1995). Formaldehyde and cancer morbidity among male employees in Denmark. *Cancer Causes Control*. 6, 354-360.

Hauptmann, M., Lubin, J. H., Stewart, P. A., (2004). Mortality from solid cancers among workers in formaldehyde industries. *Am. J. Epidemiol.* 159 (12), 1117-30.

Hayes, R. B., Raatgever, J. W., Bruyn, A., Gérin, M., (1986). Cancer of the nasal cavity and paranasal sinuses, and formaldehyde exposure. *Int. J. Cancer*. 37, 487-92.

Heck, Hd'A., Casanova-Schmitz, M., Dodd, P. B., Schachter, E. N., Witek, T., Tosun, T., (1985). Formaldehyde (CH_2O) concentrations in the blood of humans and Fischer-344 rats exposed to CH_2O under controlled conditions. *Am. Ind. Hyg. Assoc. J.* 46, 1-3.

Hong, J., Wang, D., (2014). Method for demonstrating formaldehyde removing effect of paint film with formaldehyde removing function. CN103592296 (A).

IARC Working Group on the Evaluation of Carcinogenic Risks to Humans Wood Dust and Formaldehyde. (1995). *IARC Monographs on the Evaluation of Carcinogenic Risks to Humans*. 62, 217-375. http://monographs.iarc.fr/ENG/Monographs/vol62/mono 62.pdf.

IARC Working Group on the Evaluation of Carcinogenic Risks to Humans. Formaldehyde, (2006). 2-Butoxyethanol and 1-tert-Butoxypropan-2-ol. *IARC Monographs on the Evaluation of Carcinogenic Risks to Humans*. 88, 39-287. http://monographs.iarc.fr/ ENG/Monographs/vol88/mono88-6A.pdf.

IDLH - *Documentation for Immediately Dangerous to Life or Health Concentrations*. Formaldehyde. Centers for Disease Control and Prevention. Available online: http:// www.cdc.gov/niosh/idlh/50000. HTML (accessed on 1 May 1994).

Jung, S. H., Kim, J. W., Jeon, I. G., (2001). Formaldehyde residues in formalin-treated olive flounder (Paralichthys olivaceus), black rockfish (Sebastes schlegeli) and seawater. *Aquaculture*. 194 (3-4), 253-62.

Kaszycki, P., Koloczek, H. (2002). Biodegradation of formaldehyde and its derivatives in industrial wastewater with methylotrophic yeast *Hansenula polymorpha* and with the yeast-bioaugmented activated sludge. *Biodegradation* 13, 91-9.

Kato, N., Miyawaki, N., Sakazawa, C. (1982). Oxidation of formaldehyde by resistant yeasts *Debaryomyces vanriji* and *Trichosporon penicillatum. Agric. Biol. Chem.* 46, 655-61.

Kato, N., Shirakawa, K., Kobayashi, H., (1983). The dismutation of aldehydes by a bacterial enzyme. *Agric. Biol. Chem.* 47 (1), 39-46.

Kerns, W. D., Pavkov, K. L., Donofrio, D. J., Gralla, E. J., Swenberg, J. A., (1983). Carcinogenicity of formaldehyde in rats and mice after long-term inhalation exposure. *Cancer Res.* 43, 4382-92.

Khlupova, M., Kuznetsov, B., Demkiv, O., Gonchar, M., Csöregi, E., Shleev, S., (2007a). Intact and permeabilized cells of the yeast *Hansenula polymorpha* as bioselective element for amperometric assay of formaldehyde. *Talanta.* 71 (2), 934-40.

Khlupova, M., Kuznetsov, B., Gonchar, M., Ruzgas, T., Shleev, S., (2007b). Amperometric monitoring of redox activity in intact, permeabilised, and lyophilised cells of the yeast *Hansenula polymorpha. Electrochem. Commun.* 9, 1480-5.

Kim, K. J., Jeong, M. I., Lee, D. W., Song, J. S., Kim, H. D., Yoo, E. H., Jeong, S. J., Han, S. W., Kays, S. J., Lim, Y. W., Kim, H.-H., (2010). Variation in Formaldehyde Removal Efficiency among Indoor Plant Species. *HortScience.* 45 (10), 1489-95.

Kim, W. J., Terada, N., Nomura, T., Takahashi, R., Lee, S. D., Park, J. H., Konno, A., (2002). Effect of formaldehyde on the expression of adhesion molecules in nasal microvascular endothelial cells: The role of formaldehyde in the pathogenesis of sick building syndrome. *Clin. exp. Allergy.* 32, 287-95.

Kondo, T., Morikawa, Y., Hayashi, N., (2008). Purification and characterization of alcohol oxidase from *Paecilomyces variotii* isolated as a formaldehyde-resistant fungus. *Appl. Microbiol. Biotechnol.* 77, 995-1002.

Korpan, Y. I., Gonchar, M. V., Sibirny, A. A., Martelet, C., El'skaya, A. V., Gibson, T. D., Soldatkin, A. P., (2000). Development of highly selective and stable potentiometric sensors for formaldehyde determination. *Biosens. Bioelectron.* 15, 77-83.

Korpan, Y. I., Gonchar, M. V., Starodub, N. F., Shulga, A. A., Sibirny, A. A., Elskaya, A. V., (1993). A cell biosensor specific for formaldehyde based on pH-sensitive transistors coupled to methylotrophic yeast cells with genetically adjusted metabolism. *Anal. Biochem.* 215, 216-22.

Korpan, Y. I., Soldatkin, A. P., Gonchar, M. V., Sibirny, A. A., Gibson, T. D., El'skaya, A. V., (1997). A novel enzyme biosensor spesific for formaldehyde based on pH- sensitive field effect transistors. *J. Chem. Technol. Biotechnol.* 68, 209-13.

Kovalyshyn, Y., Koval'chuk, E., Ostapovych, B., Chervinka, M., (2010). Electrochemical oxidation of formaldehyde on the platinized graphite electrode. *Visn. Lviv Univ. Ser. Chem.* 51, 342-7 (in Ukrainian).

Kumar, A., Bisht, B. S., Joshi, V. D., Dhewa, T. (2011). Review on bioremediation of polluted environment: a management tool. *Int. J. Environ. Sci.* 1, 1079-93.

Lei, L., (2014-08-27). Purifying device used for absorbing and removing formaldehyde. CN203797813 (U).

Li, J., Zhu, J., Ye, L., (2007). Determination of formaldehyde in squid by high performance liquid chromatography. *Asia Pac. J. Clin. Nutr.* 16 (1), 127-30.

Liteplo, R. G., Beauchamp, R., Meek, M. E., Chenier, R. (2002). *Formaldehyde*. Geneva. http://www.who.int/ipcs/publications/cicad/en/cicad40.pdf.

Liu, S., (2014-07-09). Geothermal type formaldehyde-removing solid wood composite floor and manufacturing method thereof. CN103912109 (A).

Liu, Y., Yang, S., (2014-08-13). Preparation method of adsorption-photocatalysis compound type formaldehyde removing coating. CN103980759 (A).

Liu, Z., Cui, T., (2014-07-16). Method for preparing natural zeolite loaded CuO nanotube composite material for removing formaldehyde. CN103920521 (A).

Lotfy, H. R., Rashed, I. G. (2002). A method for treating wastewater containing formaldehyde. *Water Res*. 36, 633-7.

Ma, C., Wang, D., Xue, W., Dou, B., Wang, H., Hao, Z., (2011). Investigation of formaldehyde oxidation over Co_3O_4-Ce_2 and Au/Co_3O_4-CeO_2 catalysts at room temperature: effective removal and determination of reaction mechanism. *Environ. Sci. Technol*. 45(8), 3628-34.

Magnusson, A., (2013). Formox and formaldehyde market update – A turbo in the future? http://www.methanolmsa.com/wp-content/uploads/2013/11/Andreas-Magnusson.pdf.

Maliutina, N. N., Taranenko, L. A., (2014). Cognitive disorders in workers engaged into formaldehyde and methanol production. *Med. Tr. Prom. Ekol*. 5, 26-31 [(in Russian].).

Mitsui, R., Kitazawa, H., Sato, T., Tanaka, M. (2006). Formaldehyde uptake by *Methylobacterium* sp. MF1 and *Acidomonas methanolica* MB 58 with the different formaldehyde assimilation pathways. *Environ. Sci*. 13, 185-92.

Monticello, T. M., Swenberg, J. A., Gross, E. A., (1996). Correlation of regional and nonlinear formaldehyde-induced nasal cancer with proliferating populations of cells. *Cancer Res*. 56, 1012-22.

Nash, T., (1953). The colorimetric estimation of formaldehyde by means of the Hantsch reaction. *Biochem. J*. 55, 416-21.

National Research Council, (1982). *Formaldehyde and Other Aldehydes* p. 2-1 to 5-96 USEPA 600/6-82-002.

Nengsih, S., Umar, A. A., Salleh, M. M., Oyama, M., (2012). Detection of formaldehyde in water: a shape-effect on the plasmonic sensing properties of the gold nanoparticles. *Sensors* (Basel). 12(8), 10309-25.

Nian, H., Meng, Q., Zhang, W., Chen, L., (2013). Overexpression of the formaldehyde dehydrogenase gene from *Brevibacillus brevis* to enhance formaldehyde tolerance and detoxification of tobacco. *Appl. Biochem. Biotech*. 169 (1), 170-180.

Nie, L., Yu, J., Li, X., Cheng, B., Liu, G., Jaroniec, M., (2013). Enhanced performance of NaOH-modified Pt/TiO_2 toward room temperature selective oxidation of formaldehyde. *Environ. Sci. Technol*. 47(6), 2777-83.

Nikitina, O., Shleev, S., Gayda, G., Demkiv, O., Gonchar, M., Gorton, L., Csöregi, E., Nistor, M., (2007). Bi-enzyme biosensor based on NAD^+- and glutathione-dependent recombinant formaldehyde dehydrogenase and diaphorase for formaldehyde assay. *Sensors and Actuators B*. 125, 1-9.

Nisnevitch, M., Sigawi, S., Smutok, O., Demkiv, O., Zakalska, O., Gayda, G., Nitzan, Y., Gonchar M., (2011). Bioconversion and detection of airborne formaldehyde by formaldehyde-metabolizing enzymes from the methylotrophic yeast *Hansenula polymorpha*. *The FEBS J*. 278, S 1, 439-440.

Offit, P., Jew, R., (2007). Addressing parents' concerns: do vaccines contain harmful preservatives, adjuvants, additives, or residuals. *Pediatrics*. 112 (6), 1394-401.

Oliveira, S. V. W. B., Moraes, E. M., Adorno, M. A. T., (2004). Formaldehyde degradation in an anaerobic packed-bed bioreactor. *Water Research*. 38 (7), 1685-94.

Omil, F., Méndez, D., Vidal, G., Méndez, R., Lema, J. M. (1999). Biodegradation of formaldehyde under anaerobic conditions. *Enzyme Microb. Tech.* 24, 255-62.

Paryzhak, S., Demkiv, O., Schuhmann, W., Gonchar, M., (2008). Intact recombinant cells of the yeast *Hansenula polymorpha*, over-producing formaldehyde dehydrogenase, as the sensitive bioelements for amperometric assay of formaldehyde. *Sensor Electronics and Microsystems Technologies* (Ukraine). 2, 28-38.

Pavlishko, H. M., Honchar, T. M., Gonchar, M. V., (2003). Chemical and enzymatic assay of formaldehyde content in fish food. Experiment. *Clin. Physiol. Biochem.* 4, 56-63 (in Ukrainian).

Pilat, P., Prokop, A., (1976). Oxidation of methanol, formaldehyde and formic acid by methanol-utilizing yeast. *Folia Microbiol.* 21 (4), 306-14.

Quievrin, G., Zhitkovich, A. (2000). Loss of DNA-protein crosslinks from formaldehyde-exposed cells occurs through spontaneous hydrolysis and an active repair process linked to proteosome function. *Carcinogenesis*. 21 (8), 1573-0.

Rehbein, H., Eichenauer, D., Feser, P., Friedrich, R., Glück, B., Harz, A., Warning, W., Werkmeister, K., Winkler, F., (1995). Formaldehyd und Dimethylamin in tiefgekuhlten Fischerzeugnissen aus dem Handeleine Bestandsaufnahme. *Archiv Lebensmittelhyg.* 46, 122-4.

Sakaguchi, K., Kurane, R., Murata, M., (1975). Assimilation of Formaldehyde and Other C1-Compounds by Gliocladium deliquescens and Paecilomyces varioti. *Agric. Biol. Chem.* 39 (9), 1695-702.

Sawicki, E., Hauser, T. R., McPherson, S., (1962). Spectrophotometric determination of formaldehyde and formaldehyde-releasing compounds with chromotropic acid, 6-amino-1-naphtol-3-sulfonic-acid (J acid), and 6-anilino-1-naphtol-3-sulfonic acid (phenyl J acid). *Anal. Chem.* 34, 1460-4.

Schmneider, C. P., Ishihara, A., Adams, T. P., Zekas, L. J., Oglesbee, M., Bertone, A. L., (2014). Analgesic effects of intraneural injection of ethyl alcohol or formaldehyde in the palmar digital nerves of horses. *Am. J. Vet. Res.* 75(9), 784-91.

Shin, S. K., Song, J. H., (2011). Modeling and simulations of the removal of formaldehyde using silver nano-particles attached to granular activated carbon. *J. Hazard. Mater.* 194, 385-92.

Shinagawa, E., Toyama, H., Matshshita, K., Tuitemwong, P., Theeragool, G., Adachi, O. (2006). A novel type of formaldehyde-oxidizing enzyme from the membrane of *Acetobacter* sp. SKU 14. *Biosci. Biotechnol. Biochem.* 70, 850-7.

Shkotova, L. V., Soldatkin, A. P., Gonchar, M. V., Schuhmann, W., Dzyadevych, S. V., (2006). Amperometric biosensor for ethanol detection based on alcohol oxidase immobilised within electrochemically deposited Resydrol film. *Materials Sci. Engineering C*. 26, 411-4.

Shleev, S. V., Shumakovich, G. P., Nikitina, O. V., Morozova, O. V., Pavlishko, H. M., Gayda, G. Z., Gonchar, M. V., (2006). Purification and characterization of alcohol oxidase from a genetically constructed over-producing strain of the methylotrophic yeast *Hansenula polymorpha*. *Biochemistry* (Moscow). 71 (3), 245-50.

Shumakovich, G. P., Shleev, S. V., Morozova, O. V., Gonchar, M. V., Yaropolov, A. I., (2007). Oxygen can be replaced by artificial electron acceptors in reactions catalyzed by alcohol oxidase. *Applied Biochem. Microbiol.* 43 (1), 19-25.

Sibirny, V. A., Gonchar, M. V., Grabek-Lejko, D., Csoregi, E., Pavlishko, H. M., Sibirny, A. A., (2008). Photometric assay of methanol and formaldehyde in industrial waste-waters using alcohol oxidase and 3-methyl-2-benzothiazolinone hydrazone. *Intern. J. Environ. Anal. Chem.* 88 (4), 289-301.

Sibirny, V., Demkiv, O., Klepach, H., Honchar, T., Gonchar, M., (2011b). Alcohol oxidase- and formaldehyde dehydrogenase-based enzymatic methods for formaldehyde assay in fish food products. *Food Chem.* 127, 774-9.

Sibirny, V., Demkiv, O., Sigawi, S., Paryzhak, S., Klepach, H., Korpan, Y., Smutok, O., Nisnevich, M., Gayda, G., Nitzan, Y., Puchalski, C., Gonchar, M., (2011a). Formaldehyde Oxidizing Enzymes and Genetically Modified Yeast *Hansenula polymorpha* Cells in Monitoring and Removal of Formaldehyde. In: *Waste Water - Evaluation and Management* (ed. F. S. G. Einschlag), ISBN 978-953-307-233-3. InTech Open Access Publisher, Rijeka, Croatia, Ch. 6, pp. 115-54.

Sibirnyi, V. A., Gonchar, M. V., Riabova, O. B., Maidan, M. M., (2005). Modern methods for formaldehyde, methanol and ethanol analysis. *Mikrobiol. Z.* (Kiev). 67 (4), 85-110.

Sigawi, S., Demkiv, O., Gayda, G., Nitzan, Y., Gonchar, M., Nisnevitch, M., (2012). Bioconversion of airborne formaldehyde by immobilized formaldehyde dehydrogenase from the recombinant methylotrophic yeast *Hansenula polymorpha*, Shanghai, China. *Proceedings of International Conference on Environmental Pollution and Public Health*, pp. 504-7.

Sigawi, S., Nisnevitch, M., Zakalska, O., Zakalsky, A., Gayda, G., Nitzan Y., Gonchar, M., (2014b). Bioconversion of airborne methylamine by immobilized methylamine oxidase from recombinant *Hansenula polymorpha. The Scientific World Journal*, 2014, Article ID 898323, 9 pages.

Sigawi, S., Smutok, O., Demkiv, O., Gayda, G., Vus, B., Nitzan, Y., Gonchar, M., Nisnevitch, M., (2014a). Detection of waterborne and airborne formaldehyde: from amperometric chemosensing to a visual biosensor based on alcohol oxidase. *Materials.* 7, 1055-68.

Sigawi, S., Smutok, O., Demkiv, O., Zakalska, O., Gayda, G., Nitzan, Y., Nisnevitch, M., Gonchar, M., (2011). Immobilized formaldehyde-metabolizing enzymes from *Hansenula polymorpha* for removal and control of airborne formaldehyde. *J. Biotechnol.* 153, 138-44.

Smutok, O., Ngounou, B., Pavlishko, H., Gayda, G., Gonchar, M., Schuhmann, W., (2006). A reagentless bienzyme amperometric biosensor based on alcohol oxidase/peroxidase and an Os-complex modified electrodeposition paint. *Sensors Actuators B: Chem.* 113 (2), 590-8.

Squire, R. A., Cameron, L. L., (1984). An analysis of potential carcinogenic risk from formaldehyde. *Regul. Toxicol. Pharmacol.* 4, 107-29.

Szende, B., Tyihak, E., Szokan, G., Katay, G., (1995). Possible role of formaldehyde in the apoptotic and mitotic effect of 1-methylascorbigen. *Pathol. Oncol. Res.* 1, 38-42.

Tian, F., Gao, W., (2014-05-21). Paint capable of removing formaldehyde. CN103804965 (A).

Tokarchuk, K. O., Zaitseva, O. V., (2014). Aldehydes participation in oxidative stress in rat thymocytes in vitro. *Ukr. Biokhim. Zh.* 86(3), 61-8.

Usanmaz, S. E., Akarsu, E. S., Vural, N., (2002). Neurotoxic effects of acute and subacute formaldehyde exposures in mice. *Environ. Toxicol. Pharmacol.* 11 (2), 93-100.

van Agteren, M. H., Keuning, S., Jansen, D. B., (1998). *Handbook on biodegradation and biological treatment of hazardous organic.* Kluwer Academic Publishers, Dordrecht, Netherlands, p. 52.

Vianello, F., Boscolo-Chio, R., Signorini, S., (2007). On-line detection of atmospheric formaldehyde by a conductometric biosensor. *Biosens. Bioelectrons.* 22 (6), 920-5.

Vidal, G., Jiang, Z. P., Omil, F., Thalasso, F., Méndez, R., Lema, J. M., (1999). Continuous anaerobic treatment of wastewaters containing formaldehyde and urea. *Bioresour. Technol.* 70 (3), 283-91.

Wang, J.-X., Shen, J., Lei, C.-S., Feng, Q., (2014). Volatile organic compound and formaldehyde emissions from Populus davidiana wood treated with low molecular weight urea-formaldehyde resin. *J. Environ. Biol.* 35(5), 989-94.

Wilton, T., Dunn, G., Eastwood, D., Minor, P. D., Martin, J., (2014). Effect of formaldehyde inactivation on poliovirus. *J. Virol.* 88(20), 11955-64.

Winter, B., Cammann, K., (1989). Formaldehyde analysis by electrochemical biosensor-FIA. *Fresenius Zeitschrift FurAnalytische Chemie.* 334 (7), 720-1.

Xu, Z., Qin, N., Wang, J., (2010). Formaldehyde biofiltration as affected by spider plant. *Bioresour. Technol.* 101 (18), 6930-4.

Yang, H., (2014-04-09). Method for removing formaldehyde in plates. CN103707373 (A).

Yoshida, K., Ishii, H., and Ishihara, Y. (2009). Bioremediation potential of formaldehyde by the marine microalga Nannochloropsis oculata ST-3 Strain. *Applied Biochem. Biotechnol.* 157 (2), 321-8.

Yu, P. H., (1998). Deamination of methylamine and angiopathy; toxicity of formaldehyde, oxidative stress and relevance to protein glycoxidation in diabetes. *J. Neural Transmis.* 52, 201-16.

Yu, P. H., Wright, S., Fan, E. H., (2003). Physiological and pathological implications of semicarbazide-sensitive amine oxidase. *Biochim. Biophys. Acta.* 1647 (1), 193-9.

Yurimoto, H., Kato, N., Sakai, Y., (2005). Assimilation, dissimilation, and detoxification of formaldehyde, a central metabolic intermediate of methylotrophic metabolism. *Chem. Rec.* 5 (6), 367-75.

Zajtseva, N. V., Zemljanova, M. A., Koldibekova, J. V., Zvezdin, V. N., Mikhajlova, D. L., Akafeva, T. J., (2014-04-20). Method for assessing negative impact of atmospheric formaldehyde on conjugation and elimination dysfunction of glutathione system in children. RU2513508 (C1).

Zhang, Y., (2014-08-13). Formaldehyde--free level laminated floor production technology. CN103978536 (A).

Zhou, P., Zhu, X., Yu, J., Xiao, W., (2013). Effects of adsorbed F, OH, and Cl ions on formaldehyde adsorption performance and mechanism of anatase TiO_2 nanosheets with exposed {001} facets. ACS Appl. *Mater. Interfaces.* 5(16), 8165-72.

In: Formaldehyde
Editor: Amy Patton

ISBN: 978-1-63482-412-5

Chapter 7

KINETICS OF THE NONBRANCHED-CHAIN ADDITION OF 1-HYDROXYALKLYL FREE RADICALS TO THE C=O BOND OF FREE UNSOLVATED FORMALDEHYDE

*Michael M. Silaev**
Department of Chemistry, Lomonosov Moscow State University, Moscow, Russia

ABSTRACT

A mechanism of the initiated nonbranched-chain process of forming 1,2-alkanediols and carbonyl compounds in alcohol–formaldehyde systems is suggested. The quasi-steady-state treatment is used to obtain kinetic equations that can describe the nonmonotonic (with a maximum) dependences of the formation rates of the products on the concentration of free unsolvated formaldehyde. An empirical equation for calculating the concentration of the free unsolvated form of formaldehyde in alcohol–formaldehyde (including water/ethanediol–formaldehyde) systems at various temperatures and total formaldehyde concentrations and an equation for evaluating solvent concentrations in these systems were derived.

Keywords: nonbranched-chain process, free formaldehyde, 1-hydroxyalkyl radical, formyl radical, competing reaction, equation.

INTRODUCTION

Free radicals add to the carbon atom at the double bond of the carbonyl group of dissolved free (unsolvated, monomer) formaldehyde. The concentration of free formaldehyde in the solution at room temperature is a fraction of a percent of the total formaldehyde concentration, which includes formaldehyde chemically bound to the solvent [1]. The

* Corresponding Author: mmsilaev@rc.chem.msu.ru.

concentration of free formaldehyde exponentially increases with increasing temperature [2]. The energy released as a result of this addition, when the C=O bond is converted into an ordinary bond, is 30 to 60 kJ mol^{-1} (according to the data on the addition of C_1–C_4 alkyl radicals in the gas phase under standard conditions [3–6]). The resulting free 1:1 adduct radicals can both abstract hydrogen atoms from the nearest-neighbor molecules of the solvent or unsolvated formaldehyde and, due to its structure, decompose by a monomolecular mechanism including isomerization [7, 8].

ADDITION OF 1-HYDROXYALKLYL FREE RADICALS WITH TWO OR MORE CARBON ATOMS

Free 1-hydroxyalkyl radicals (which result from the abstraction of a hydrogen atom from the carbon atom bonded to the hydroxyl group in molecules of saturated aliphatic alcohols but methanol under the action of chemical initiators [9, 10], light [11, 12], or ionizing radiation [13, 14]) add at the double bond of free formaldehyde dissolved in the alcohol, forming 1,2-alkanediols [7–10, 12–18], carbonyl compounds, and methanol [14, 15] *via* the chaining mechanism. (The yields of the latter two products in the temperature range of 303 to 448 K are one order of magnitude lower). In these processes, the determining role in the reactivity of the alcohols can be played by the desolvation of formaldehyde in alcohol–formaldehyde solutions, which depends both on the temperature and on the polarity of the solvent [2, 14]. For the γ-radiolysis of 1(or 2)-propanol–formaldehyde system at a constant temperature, the dependences of the radiation-chemical yields of 1,2-alkanediols and carbonyl compounds as a function of the formaldehyde concentration show maxima and are symbatic [13, 15]. For a constant total formaldehyde concentration of 1 mol dm^{-3}, the dependence of the 1,2-alkanediol yields as a function of temperature for 303–473 K shows a maximum, whereas the yields of carbonyl compounds and methanol increase monotonically [14] (along with the concentration of free formaldehyde [2]). In addition to the above products, the nonchain mechanism in the γ-radiolysis of the solutions of formaldehyde in ethanol and 1- and 2-propanol gives ethanediol, carbon monoxide, and hydrogen in low radiation-chemical yields (which, however, exceed the yields of the same products in the γ-radiolysis of individual alcohols) [7, 14, 15]. The available experimental data can be described in terms of the following scheme of reactions.

Scheme

Chain initiation

1. $\mathrm{I} \xrightarrow{2k_1} 2\,\mathrm{R}_0^{\bullet}$;

1a. $\mathrm{R}_0^{\bullet} + \mathrm{ROH} \xrightarrow{k_{1a}} \mathrm{ROH} + {}^{\bullet}\mathrm{R}_{(-\mathrm{H})}\mathrm{OH}$.

Chain propagation

2. ${}^{\bullet}\mathrm{R}_{(-\mathrm{H})}\mathrm{OH} + \mathrm{CH_2O} \xrightarrow{k_2} \mathrm{R}_{(-\mathrm{H})}(\mathrm{OH})\mathrm{CH_2O}^{\bullet}$;

3. $\mathrm{R}_{(-\mathrm{H})}(\mathrm{OH})\mathrm{CH_2O}^{\bullet} + \mathrm{ROH} \xrightarrow{k_3} \mathrm{R}_{(-\mathrm{H})}(\mathrm{OH})\mathrm{CH_2OH}$ +

$$^{\bullet}R_{(-H)}O;$$

$$3a.\ R_{(-H)}(OH)CH_2O^{\bullet} \xrightarrow{k_{3a}} R_{(-2H)}HO + {}^{\bullet}CH_2OH$$
$$(\text{or } R'R''CO + {}^{\bullet}CH_2OH);$$

$$3b.\ {}^{\bullet}CH_2OH + ROH \xrightarrow{k_{3b}} CH_3OH + {}^{\bullet}R_{(-H)}OH.$$

Inhibition

$$4.\ R_{(-H)}(OH)CH_2O^{\bullet} + CH_2O \xrightarrow{k_4} R_{(-H)}(OH)CH_2OH + {}^{\bullet}CHO.$$

Chain termination

$$5.\ 2{}^{\bullet}R_{(-H)}OH \xrightarrow{2k_5} R_{(-H)}(OH)R_{(-H)}OH$$
$$(\text{or: } ROH + R_{(-2H)}HO,$$
$$ROH + R'R''CO);$$

$$6.\ {}^{\bullet}R_{(-H)}OH + {}^{\bullet}CHO \xrightarrow{k_6} R_{(-H)}(OH)CHO$$
$$(\text{or: } R_{(-2H)}HO + CH_2O,$$
$$R'R''CO + CH_2O,$$
$$ROH + CO);$$

$$7.\ 2{}^{\bullet}CHO \xrightarrow{2k_7} HC(O)CHO$$
$$(\text{or: } CH_2O + CO,$$
$$2CO + H_2).$$

In these reactions, I is an initiator, e.g., a peroxide [9, 10]; $R_0^{\bullet}$, some reactive radical (initiator radical); R, an alkyl; ROH, a saturated aliphatic alcohol, either primary or secondary, beginning from ethanol; CH_2O, the unsaturated molecule – free formaldehyde; ${}^{\bullet}CH_2OH$, the 1-hydroxymetyl fragment radical; ${}^{\bullet}R_{(-H)}OH$, the reactive 1-hydroxyalkyl addend radical, beginning from 1-hydroxyethyl; $R_{(-H)}(OH)CH_2O^{\bullet}$, the reactive hydroxyalkoxyl 1:1 adduct radical; ${}^{\bullet}CHO$, the low-reactive formyl radical (inhibitor); R_0H, the molecular product; $R_{(-H)}(OH)CH_2OH$, 1,2-alkanediol; $R_{(-2H)}HO$, an aldehyde in the case of a primary alcohol and an R'R"CO ketone in the case of a secondary alcohol; $R_{(-H)}(OH)R_{(-H)}OH$, a vicinal alkanediol; $R_{(-H)}(OH)CHO$, a hydroxyaldehyde. The chain evolution stage of *Scheme* includes consecutive reaction pairs 2–3, 2–3a, and 3a–3b; parallel (competing) reaction pairs 3–3a, 3–3b, 3–4, and 3a–4; and consecutive–parallel reactions 2 and 4.

In addition, it seems unlikely that free adduct radicals will add to formaldehyde at higher temperatures the reaction of adding is unlikely because this would result in an ether bond. The addition of hydroxymethyl radicals to formaldehyde, which is in competition with reaction 3b, is not included as well, because there is no chain formation of ethanediol at 303–448 K [14]. At the same time, small amounts of ethanediol can form *via* the dimerization of a small fraction of hydroxymethyl radicals, but this cannot have any appreciable effect on the overall process kinetics. The addition of free formyl radicals to formaldehyde cannot proceed at a significant rate, as is indicated by the fact that there is no chain formation of glycol aldehyde in the systems examined [14].

The mechanism of the decomposition of the free adduct radical *via* reaction 3a, which includes the formation of an intramolecular H···O bond and isomerization, can be represented as follows [7, 8, 15]:

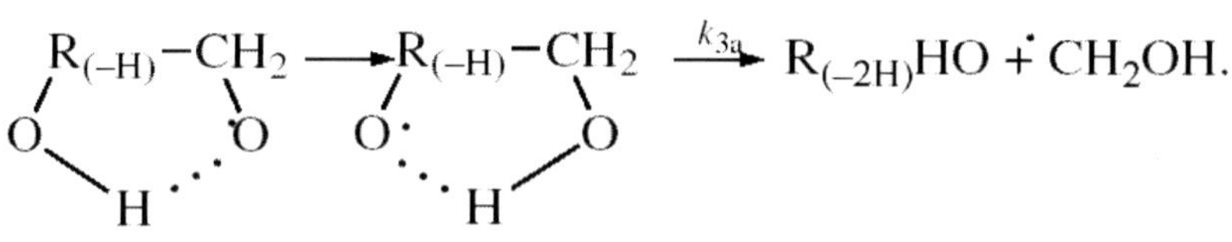

The probability of the occurrence of reaction 3a should increase with increasing temperature. This is indicated by experimental data presented above [7, 8, 15]. The decomposition of the hydroxyalkoxyl radical. $R_{(-H)}(OH)CH_2O^{\bullet}$ (reaction 3a) is likely endothermic. The endothermic nature of reaction 3a is indirectly indicated by the fact that the decomposition of simple C_2–C_4 alkoxyl radicals $RO^{\bullet}$ in the gas phase is accompanied by heat absorption: (ΔH°_{298} = 30–90 kJ mol^{-1} [4–6]). Reaction 3b, subsequent to reaction 3a, is exothermic, and its heat for C_2–C_3 alcohols in the gas phase is ΔH°_{298} = –40 to –60 kJ mol^{-1} [4–6]. As follows from the above scheme of the process, reactions 3a and 3b, in which the formation and consumption of the highly reactive free radical hydroxymethyl take place (at equal rates under steady-state conditions), can be represented as a single bimolecular reaction 3a,b occurring in a "cage" of solvent molecules.

The free formyl radical resulting from reaction 4, which is in competition with reactions 3 and 3a, is comparatively low-reactive because its spin density can be partially delocalized from the carbon atom *via* the double bond toward the oxygen atom, which possesses a higher electron affinity [3]. For example, in contrast to the methyl and alkoxyl π-radicals, the formyl σ-radical can be stabilized in glassy alcohols at 77 K [19]. In the gas phase, the dissociation energy of the C–H bond in formyl radicals is half that for acetyl radicals and is about 5 times lower than the dissociation energy of the C_α–H bond in saturated C_1–C_3 alcohols [3].

As distinct from reactions 3 and 3a,b, reaction 4 leads to an inefficient consumption of hydroxyalkoxyl adduct radicals, without regenerating the initial 1-hydroxyalkyl addend radicals. Reaction 4 together with reaction 6 (mutual annihilation of free formyl and chain-carrier 1-hydroxyalkyl radicals) causes the inhibition of the nonbranched-chain process. For the disproportionation of the free radicals, the heats of reactions 5–7 for C_1–C_3 alcohols in the gas phase vary in the range of ΔH°_{298} = –135 to –385 kJ mol^{-1} [3–6].

The rates of the chain formation of 1,2-alkanediols in reaction 3 (and their nonchain formation in reaction 4), carbonyl compounds in reaction 3a, and methanol in reaction 3b are given by the following equations[1]:

$$V_{3,4}(R_{(-H)}(OH)CH_2OH) = \frac{V_1(\alpha l + x)k_2 x}{k_2 x^2 + (\alpha l + \beta + x)\sqrt{2k_5 V_1}}, \qquad (1)$$

[1] The rate equations were derived by quasi-steady-state treatment, which is most suitable for describing the processes including at least eight to ten reactions with four to six different free radicals and at most three to seven experimental points in their functional curves, using the condition for the first steps of the process that makes it possible to reduce the exponent of term $2k_5[^{\bullet}R_{(-H)}OH]^2$ to 1 in equation $d[R^{\bullet}]/dt = 0$ [7, 15]: $k_6 = \sqrt{2k_5 2k_7}$ [20] and $V_1 = V_5 + 2V_6 + V_7 = (\sqrt{2k_5}[^{\bullet}R_{(-H)}OH] + \sqrt{2k_7}[^{\bullet}CHO])^2$. The relations between the reaction rates and radiation-chemical yields for radiation-chemical processes are reported by Silaev [21].

$$V_{3a}(R_{(-2H)}HO) = V_{3b}(CH_3OH) = \frac{V_1 \beta k_2 x}{k_2 x^2 + (\alpha l + \beta + x)\sqrt{2k_5 V_1}}, \quad (2)$$

where V_1 is the initiation rate, l is the molar concentration of the saturated alcohol at a given total concentration of formaldehyde[2] dissolved in it, x is the molar concentration of free formaldehyde ($l >> x$), k_2 is the rate constant of reaction 2 (addition of 1-hydroxyalkyl free radical to free formaldehyde), and $\alpha = k_3/k_4$ and $\beta = k_{3a}/k_4$ (mol dm^{-3}) are the ratios of the rate constants of the competing (parallel) reactions. Estimates of $2k_5$ were reported by Silaev *et al.* [22, 24]. From the extremum condition for the reaction 3a rate function, $\partial V_{3a} / \partial x = 0$, we derived the following analytical expression: $k_2 = (\alpha l_m + \beta)\sqrt{2k_5 V_1} / x_m^2$.

Equations (1) and (2) taken on the condition that $k_2 x^2 >> (\alpha l + x)\sqrt{2k_5 V_1}$ (the descending branch of the curve having a maximum) can be transformed to equations that can be used for a preliminary estimation of parameters α and β:

$$V_{3,4} = (V_1 / \varphi)[(\alpha l / x) + 1], \quad (1a)$$

$$V_{3a} = V_{3b} = V_1 \beta / \varphi x, \quad (2a)$$

where $\varphi = 2$ for the maximum (when $k_2 x^2 \cong (\alpha l + x)\sqrt{2k_5 V_1}$) and $\varphi = 1$ for the descending portion of the curve.

The overall process rate is a complicated function of the formation and disappearance rates of the $^{\bullet}R_{(-H)}OH$ and $^{\bullet}CHO$ free radicals:

$$V(R_{(-H)}(OH)CH_2OH, R_{(-2H)}HO, CH_3OH) = V_{1a} + V_3 + V_{3b} - V_4 - V_5 + V_7. \quad (3)$$

The ratios of the rates of the competing reactions are $V_3/V_4 = \alpha l/x$ and $V_{3a}/V_4 = \beta/x$, and the chain length is $\nu = (V_3 + V_{3a})/V_1$. The ratio of the rates of formation of 1,2-alkanediol and the carbonyl compound is a simple linear function of x:

$$V_{3,4}(R_{(-H)}(OH)CH_2OH) / V_{3a}(R_{(-2H)}HO) = (k_4 / k_{3a})\, x + (k_3 / k_{3a}) l. \quad (4)$$

The equations for the rates of chain-termination reactions 5–7 are given by the following equations:

$$V_5 = V_1^2 2k_5 (\alpha l + \beta + x)^2 / f^2, \quad (5)$$

[2] The alcohol concentration in alcohol–formaldehyde solutions at any temperature can be estimated by the method suggested in [22, 23]. The data necessary for estimating the concentration of free formaldehyde using the total formaldehyde concentration in the solution are reported by Silaev *et al.* [2, 22].

$$2V_6 = 2V_1\sqrt{2k_5V_1}\left(\alpha l+\beta+x\right)k_2x^2/f^2, \tag{6}$$

$$V_7 = V_1\left(k_2x^2\right)^2/f^2, \tag{7}$$

where $f = k_2x^2 + (\alpha l + \beta + x)\sqrt{2k_5V_1}$.

Neutral formaldehyde solutions in alcohols at room temperature primarily consist of a mixture of formaldehyde polymer solvates reversibly bound to alcohols; these polymer solvates differ in molecular mass and have the general formula $RO(CH_2O)_nH$, where n = 1–4 [1]. The concentration of formaldehyde that occurs in solution as a free, unsolvated active species chemically unbound to the solvent (this species is capable of scavenging free radicals) at room temperature is lower than a percent of the total formaldehyde concentration [1]. The concentration of the free formaldehyde species in solutions was determined by high-temperature spectrophotometry in the range 335–438 K at total formaldehyde concentration of polymer solvates) of 1.0–8.4 mol dm^{-3} in water, ethanediol, methanol, ethanol, 1-propanol, 2-propanol, and 2-methyl-2-propanol [2]. This concentration increases with temperature according to an exponential law, and it can be as high as a few percent of the total concentration in solution under the test conditions, up to 19.3% in the case of 2-methyl-2-propanol at a total concentration of 1.0 mol dm^{-3} and a temperature of 398 K. The following empirical equation relating the concentration x (mol dm^{-3}) of free formaldehyde to temperature T (K) and the total concentration c_0 in solution (measured at room temperature), was developed by the treatment of 101 data points [2, 22]:

$$\lg x = -a\,(10^3/T) + b + h\,\lg c_0, \tag{8}$$

where the coefficients a and b were calculated as the parameters of a straight-line equation by the least-squares technique from the dependence of lg x on $1/T$ at c_0 = 1.0 mol dm^{-3} for various solvents, and the coefficient h was obtained as the average value of the slopes of lg x as linear functions of lg c_0 at various series of fixed temperatures. The Table summarizes these coefficients for each solvent. As regards the experimental data, the error in the calculations of the concentration x of free formaldehyde made by Eq. (8) in the specified temperature range was no higher than 25%.

Table. Coefficients of empirical Eq. (8) for the estimation of the concentration *x* of the free formaldehyde species in polar solvent–formaldehyde systems [2, 22]

Solvent	Coefficient		
	a	b	h
Water	2.36	4.45	0.80
Ethanediol	1.83	2.60	1.28
Methanol	3.11	5.58	$0.22\,c_0/\lg c_0$
Ethanol	3.10	5.92	$1.10\,(10^3/T) - 1.44$
1-Propanol	2.42	4.47	1.30
2-Propanol	2.42	4.64	1.05
2-Methyl-2-propanol	3.19	7.31	0.96

On the assumption that the dependence of the density of a given solution on the concentration of formaldehyde is similar to the analogous linear dependence found for aqueous formaldehyde solutions (0–14 mol dm^{-3}; 291 K) [25], the concentrations l_T (mol dm^{-3}) of alcohols in alcohol–formaldehyde solutions at a certain temperature can be estimated by the equation

$$l_T = \frac{(10^3 d - 21.6 c_0) d_T}{(d + 8.4 \times 10^{-3} c_0) M}, \tag{9}$$

where c_0 is the total formaldehyde concentration (mol dm^{-3}); M is the molecular mass (g mol^{-1}) of the solvent; d and d_T are the solvent densities (g cm^{-3}) at room and given temperatures, respectively; the coefficients 8.4×10^{-3} and 21.6 have the units of 10^3g mol^{-1} and g mol^{-1}, respectively.

Earlier [2], it was found that the concentration of the free formaldehyde species decreased with the solvent permittivity at a constant temperature. Water is an exception. Although water is more polar than alcohols, the concentration of formaldehyde in an aqueous solution is anomalously high and reaches the level of its concentration in 2-propanol, all other factors being the same (see Fig. 1) [2, 22]. This can be due to the specific instability of hydrated formaldehyde species and the ease of their conversion into free formaldehyde with increasing temperature.

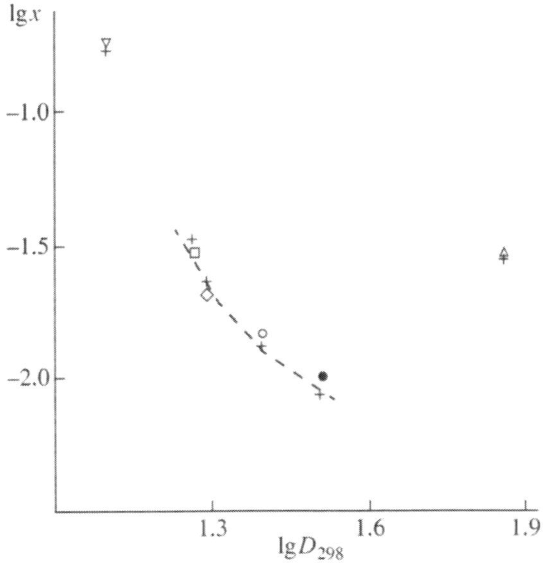

Figure 1. Logarithmic plot of the experimental concentrations x (mol dm^{-3}) of free formaldehyde at its total concentration c_0 = 1.0 mol dm^{-3} and 395 ± 3 K in (△) water, (●) methanol, (○) ethanol, (◇) 1-propanol, (□) 2-propanol, and (▽) 2-methyl-2-propanol as functions of the permittivity D_{298} of these solvents (+ refers to the concentrations x in the above solvents calculated by empirical Eq. (8).

As an example, the Fig. 2 illustrates the use of Eqs. (1) and (2) for describing the experimental dependences of the formation rates of 1,2-butanediol (curve *1*) in reactions 3 and 4 and propanal (curve *2*) in reaction 3a on the concentration of free formaldehyde in the

1-propanol–formaldehyde reacting system at a total formaldehyde concentration of 2.0 to 9.5 mol dm^{-3} and temperature of 413 K [7, 15, 26].

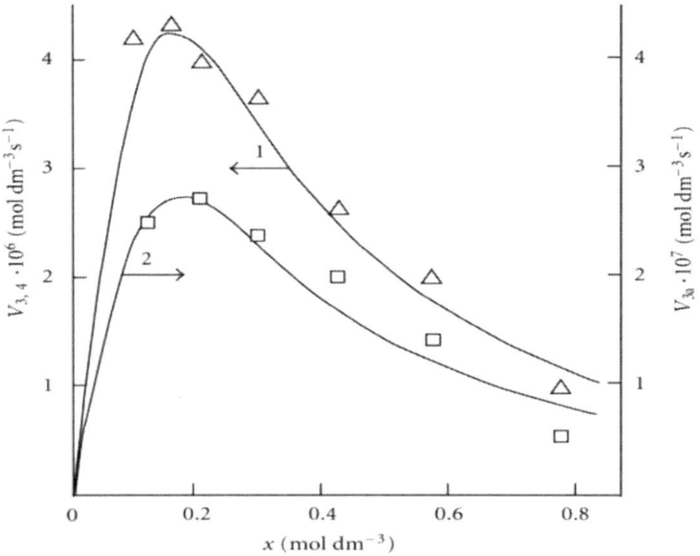

Figure 2. Reconstruction of the functional dependence (curves) of the product formation rates $V_{3,4}$ and V_{3a} on the concentration x of free formaldehyde (model optimization with respect to the parameters α, β and k_2) from empirical data (symbols) for the 1-propanol–formaldehyde system at 413 K [7, 15, 26]: (*1*, △) calculation using Eq. (1), the product is 1,2-butanediol, standard deviation of $S_Y = 2.20 \times 10^{-7}$; (*2*, □) calculation using Eq. (2), the product is propanal, $S_Y = 2.38 \times 10^{-8}$.

The concentration dependence of the propanal formation rate was described using the estimates of kinetic parameters obtained for the same dependence of the 1,2-butanediol formation rate. We considered these data more reliable for the reason that the carbonyl compounds forming in the alcohol–formaldehyde systems can react with the alcohol and this reaction depends considerably on the temperature and acidity of the medium [1]. The mathematical modeling of the process was carried out using a ^{137}Cs γ-radiation dose rate of $P = 0.8$ Gy s^{-1} [13, 26], a total initiation yield of $G(CH_3CH_2\dot{C}HOH) = 9.0$ particles per 100 eV [7, 15] ($V_1 = 4.07 \times 10^{-7}$ mol dm^{-3} s^{-1}), and $2k_5 = 4.7 \times 10^9$ dm^3 mol^{-1} s^{-1}. The following values of the parameters were obtained: $\alpha = 0.36 \pm 0.07$, $\beta = 0.25 \pm 0.05$ mol dm^{-3}, and $k_2 = (6.0 \pm 1.4) \times 10^3$ dm^3 mol^{-1} s^{-1}.

Note that, as compared to the yields of 1,2-propanediol in the γ-radiolysis of the ethanol–formaldehyde system, the yields of 2,3-butanediol in the γ-radiolysis of the ethanol–acetaldehyde system are one order of magnitude lower [26]. Using data from [7, 15], it can be demonstrated that, at 433 K, the double bond of 2-propen-1-ol accepts the 1-hydroxyethyl radical 3.4 times more efficiently than the double bond of formaldehyde [27].

Addition of Hydroxymethyl Free Radicals

The addition of hydroxymethyl radicals to the carbon atom at the double bond of free formaldehyde molecules in methanol, initiated by the free-radical mechanism, results in the chain formation of ethanediol [16]. In this case, reaction 3a in *Scheme* is the reverse of reaction 2, the 1-hydroxyalkyl radical $^{\bullet}R_{(-H)}OH$ is the hydroxymethyl radical $^{\bullet}CH_2OH$, so reaction 3b is eliminated($k_{3b} = 0$), and reaction 5 yields an additional amount of ethanediol *via* the dimerization of chain-carrier hydroxymethyl radicals (their disproportionation can practically be ignored [28]). The scheme of these reactions is presented in [17].

The rate equation for ethanediol formation by the chain mechanism in reaction 3 and by the nonchain mechanism in reactions 4 and 5 in the methanol–formaldehyde system has a complicated form[3] as compared to Eq. (1) for the formation rate of the other 1,2-alkanediols [8]:

$$V_{3,4,5}(CH_2OH)_2 =$$
$$V_1[f(\alpha l + x)k_2 x + V_1 2k_5(\alpha l + \beta + x)^2]/f^2, \quad (10)$$

where $f = k_2x^2 + (\alpha l + \beta + x)\sqrt{2k_5V_1}$.

Equation (10) when $k_2x^2 << (\alpha l + \beta + x)\sqrt{2k_5V_1}$ (ascending branch of the curve having a maximum) and $\alpha l >> \beta$ (practically without reaction 3a) is transformed to a simple directly-proportional dependence on the concentration x of free formaldehyde, which can be used to pre-estimate the parameter k_2:

$$V_{3,4,5} = (V_1/\varphi^2)[(\varphi\, k_2\, x / \sqrt{2k_5V_1}\,) + 1], \quad (10a)$$

where $\varphi = 1$ for the ascending portion of the curve and $\varphi = 2$ for the maximum, when $k_2x^2 \cong (\alpha l + \beta + x)\sqrt{2k_5V_1}$.

If the rate of ethanediol formation by the dimerization mechanism in reaction 5 is ignored for the reason that it is small as compared to the total rate of ethanediol formation in reactions 3 and 4, Eq. (10) will be identical to Eq. (1). After the numerator and denominator on the right-hand side of Eq. (1) are divided by $k_{-2} \equiv k_{3a}$, one can replace k_2 in this equation with $K_2 = k_2/k_{-2}$, which is the equilibrium constant for the reverse of reaction 2. Ignoring the reverse step of reaction 2 ($k_{3a} = 0, \beta = 0$) can further simplify Eq. (1). In this case, the rate constant k_2 is effective.

REFERENCES

[1] S. K. Ogorodnikov, Formal'degid (Formaldehyde), *Khimiya,* Leningrad, 1984.

[2] M. M. Silaev, A. V. Rudnev, and E. P. Kalyazin, "Formaldehyde. III. Concentration of Free Formaldehyde as a Function of Temperature, Polarity of Solvents, and Total Concentration of Formaldehyde in Solution", *Zhurnal Fizicheskoi Khimii,* 1979, vol. 53, no. 7, pp. 1647–1651.

[3] L. V. Gurvich, G. V. Karachevtsev, V. N. Kondrat'ev, Yu. A. Lebedev, V. A. Medvedev, V. K. Potapov, and Yu. S. Khodeev, "Energii razryva khimicheskikh

[3]In an earlier publication [15], this equation does not take into account reaction 3a.

svyazei. Potentsialy ionizatsii i srodstvo k elektronu"("Bond Dissociation Energies, Ionization Potentials, and Electron Affinity"), V. N. Kondrat'ev, Editor, *Nauka,* Moscow, 1974.
[4] S. W. Benson, "Thermochemical Kinetics: Methods for the Estimation of Thermochemical Data and Rate Parameters", 2nd Edition, *Wiley,* New York, 1976.
[5] J. B. Pedley, R. D. Naylor, and S. P. Kirby, "Thermochemical Data of Organic Compounds", 2nd Edition, Chapman & Hall, London, 1986.
[6] Yu. D. Orlov, Yu. A. Lebedev, and I. Sh. Saifullin, "Termokhimiya organicheskikh svobodnykh radikalov" ("Thermochemistry of Organic Free Radicals"), A. M. Kutepov, Editor, *Nauka,* Moscow, 2001.
[7] M. M. Silaev and L. T. Bugaenko, "Kinetics of the Addition of α-Hydroxyalkyl Radicals to 2-Propen-1-ol and Formaldehyde", *Kinetics and Katalysis,* 1994, vol. 35, no. 4, pp. 509–513, English Translation in: Kinetics and Katalysis, 1994, vol. 35, no. 4, 1994, pp. 463–467.
[8] M. M. Silaev, "Simulation of Nonbranched Chain Processes for Producing 1,2-Alkanediols in Alcohol–Formaldehyde Systems", *Teoreticheskie Osnovy Khimicheskoi Tekhnologii,* 2007, vol. 41, no. 4, pp. 379–384, English Translation in: Thoretical Foundations Chemical Engineering, 2007, vol. 41, no. 4, pp. 357–361.
[9] Oyama, M., "A Free-Radical Reaction of Primary and Secondary Alcohols with Formaldehyde", *The Journal of Organic Chemistry,* 1965, vol. 30, no. 7, pp. 2429–2432.
[10] G. I. Nikitin, D. Lefor, and E. D. Vorob'ev, "Free Radical Reaction of Primary Alcohols with Formaldehyde", *Izvestiya Akademii Nauk SSSR, Ser. Khimiya,* 1966, no. 7, pp. 1271–1272.
[11] W. H. Urry, F. W. Stacey, E. S. Huyser, and O. O. Juveland, "The Peroxide- and Light-Induced Additions of Alcohols to Olefins", *Journal of the American Chemical Society,* 1954, vol. 76, no. 2, pp. 450–455.
[12] M. B. Dzhurinskaya, A. V. Rudnev, and E. P. Kalyazin, "High Temperature UV Photolysis of Formaldehyde in Liquid Methanol", Vestnik Moskovskogo Universiteta, Ser. 2: *Khimiya,* 1984, vol. 25, no. 2, pp. 173–176.
[13] E. P. Kalyazin, E. P. Petryaev, and O. I Shadyro, "Reaction between Oxyalkyl Radicals and Aldehydes", *Zhurnal Organicheskoi Khimii,* 1977, vol. 13, no. 2, pp. 293–295.
[14] A. I. Novoselov, A.I., Silaev, M.M., and L. T. Bugaenko, "Effect of Temperature on the Yields of Final Products in the γ-Radiolysis of Formaldehyde Solutions in C_1–C_3 Alkanols", *Khimiya Vysokich Energii,* 2004, vol. 38, no. 4, pp. 270–272, English Translation in: High Energy Chemistry, 2004, vol. 38, no. 4, 2004, pp. 236–238.
[15] M. M. Silaev and L. T. Bugaenko, "Mathematical Simulation of the Kinetics of Radiation Induced Hydroxyalkylation of Aliphatic Saturated Alcohols", *Radiation Physics and Chemistry,* 1992, vol. 40, no. 1, pp. 1–10.
[16] A. I. Novoselov, M. M. Silaev, and L. T. Bugaenko, "Dependence of Ethanediol Yield on Formaldehyde Concentration in γ-Radiolysis of Methanol–Formaldehyde System at 373–473 K", *Khimiya Vysokikh Energii,* 2008, vol. 42, no. 1, pp. 74–75, English Translation in: High Energy Chemistry, 2008, vol. 42, no. 1, pp. 69–70.
[17] A. I. Novoselov, M. M. Silaev, and L. T. Bugaenko, "γ-Induced Single-Step Synthesis of Ethylene Glycol from Methanol–Formaldehyde Solution", *Theoreticheskie Osnovy*

Khimicheskoy Tekhnologii, 2010, vol. 44, no. 4, pp. 450–453, English Translation in: Theoretical Foundation of Chemical Engineering, 2010, vol. 44, no. 4, pp. 432–435.

[18] A. I. Novoselov, M. M. Silaev, and L. T. Bugaenko, "Dependence of 1,2-Propanediol Yield on Formaldehyde Concentration in γ-Radiolysis of Ethanol–Formaldehyde System at 373–473 K", *Khimiya Vysokikh Energii,* 2007, vol. 41, no. 1, p. 58, English Translation in: High Energy Chemistry, 2007, vol. 41, no. 1, p. 53.

[19] S. Ya. Pshezhetskii, A. G. Kotov, V. K. Milinchuk, V. A. Roginskii, and V. I. Tupikov, "EPR svobodnykh radikalov v radiatsionnoi khimii" ("ESR of Free Radicals in Radiation Chemistry"), *Khimiya,* Moscow, 1972.

[20] L. Bateman, "Olefin Oxidation", *Quarterly Reviews,* 1954, vol. 8, no. 2, pp. 147–167.

[21] M. M. Silaev, "Simulation of the Nonbranched-Chain Addition of Saturated Free Radicals to Alkenes and Their Derivatives Yielding 1:1 Adducts", *Teoreticheskie Osnovy Khimicheskoi Tekhnologii,* 2007, vol. 41, no. 3, pp. 280–295, English Translation in: Theoretical Foundations of Chemical Engineering, 2007, vol. 41, no. 3, pp. 273–278.

[22] M. M. Silaev, "Applied Aspects of the γ-Radiolysis of C_1–C_4 Alcohols and Binary Mixtures on Their Basis", *Khimiya Vysokikh Energii,* 2002, vol. 36, no. 2, pp. 97–101, English Translation in: High Energy Chemistry, 2002, vol. 36, no. 2, pp. 70–74.

[23] M. M. Silaev, "Estimating the Solvent Concentration in Formaldehyde Solutions at Various Temperatures", *Zhurnal Fizicheskoy Khimii*, 1993, vol. 67, no. 9, p. 1944.

[24] M. M. Silaev, L. T. Bugaenko, and E. P. Kalyazin, "On the Possibility of Adequately Estimating of the Rate Constants for the Reaction of Hydroxyalkyl Radicals with Each Other Using the Self-Diffusion Coefficients or Viscosities of the Corresponding Alcohols", *Vestnik Moskovskogo. Univiversiteta, Ser. 2: Khimiya,* 1986, vol. 27, no. 4, pp. 386–389.

[25] J. F. Walker, Formaldehyde, Reinhold, New York, 1953, English Translation under the title Formal'degid, *Goskhimizdat,* Moscow, 1957, p. 106.

[26] O. I. Shadyro, "Radiation-chemical Conversions of Aldehydes in Various Systems", Ph.D. Thesis (Chemistry), Belarusian State University, Minsk, 1975.

[27] M. M. Silaev, "Relative Reactivity of α-Hydroxyethyl Radicals for 2-Propene-1-ol and Formaldehyde Double-Bond Addition", Vestnik Moskovskogo Universiteta, Ser. 2: *Khimiya,* 1993, vol. 34, no. 3, p. 311.

[28] H. Seki, R. Nagai, and M. Imamura, "γ-Radiolysis of a Binary Mixture of Methanol and Water. The Formation of Formaldehyde in the Radiolysis of Liquid Methanol", *Bulletin of the Chemical Society of Japan,* 1968, vol. 41, no. 12, pp. 2877–2881.

In: Formaldehyde
Editor: Amy Patton
ISBN: 978-1-63482-412-5

Chapter 8

MECHANOCHEMICAL SYNTHESIS OF MOLYBDATES: INFLUENCE OF MECHANICAL ACTIVATION ON STRUCTURE OF MOLYBDENUM OXIDES AND IRON

A. A. Il'in, R. N. Rumyantsev, A. P. Il'in and N. N. Katkova
Ivanovo State University of Chemistry and Technology; Ivanovo, Russia

ABSTRACT

In this work we established influence of mechanical activation on structure and properties oxides of iron and molybdenum. The process of mechanochemical synthesis of molybdates of iron and calcium have been studied, and also their catalytic properties in reaction of partial oxidation of methanol in formaldehyde. It is established that 2% addition in structure of the catalyst of calcium molybdate increases durability of contact weight without essential change of catalytic activity.

Owing to method of the laser analysis it has been established that original MoO_3 consists of large particles up to 120 microns in size (figure 1). Fraction of particles less than 1 micron is about 1%. In special equipment the mechanical activation which takes about 15 mins helps to reduce the particle size to 10 microns and fraction of particles less than 1 micron is 50%. After increasing time of activation to 45+ mins., the particles will aggregate.

It happens because owing to dispersion with the mechanical influence particles achieve their minimum size and the process almost stops, after this the process of secondary aggregation begins and we can see there how excess energy reduce.

Microstructure of objects forms by 2 different processes: reducing the size of individual particles during their destruction and formation of these particles with sufficiently large combinations. Duration of the mechanical activation has a big influence on surface of samples. Original MoO_3 has pretty low surface area, about 1,5 m^2/g. After 45 mins of mechanical activation surface area increases up to 55 m^2/g. If we continue to work with it, its size reduces to 18 m^2/g after 90 mins of mechanical activation. It happens because these particles combine so close to each other so inner surface of pores do not allow to nitrogen and other gases to absorb (table 1).

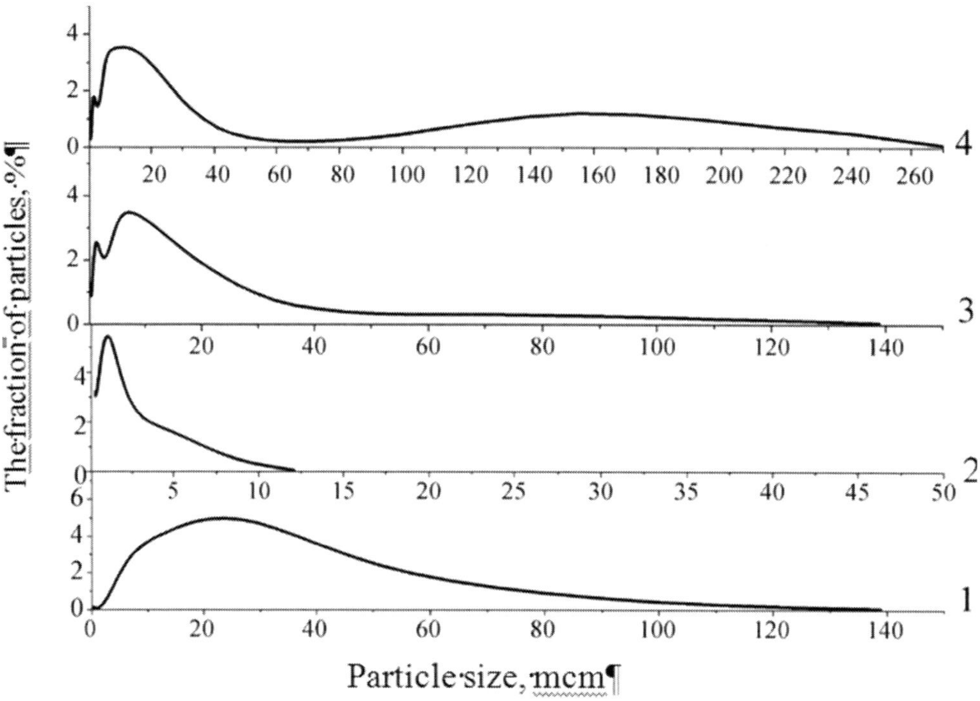

1 – original MoO_3 2-15 mins of MA 3 – 45 mins of MA 4-90 mins of MA.

Figure 1. Results of laser analysis of molybdenum oxide powder.

Table 1. Speed of change in physical and chemical properties of MoO_3 during mechanical activation

Parameter	Time of MA (min)					
	0	15	30	45	60	90
Surface Area (m^2/g)	1,5	15	30	55	24	18
Quantity of Fe (%)	-	0,1	0,19	0,23	0,31	0,38
Area of coherent dispersion, (Å)	340	310	280	270	250	220
Quantity of micro-deformations (%)	0,26	0,45	0,52	0,55	0,56	0,57

During mechanical activation of different materials the mechanical influence changes not only the crushed substance, but also the grinding bodies and a working surface of a mill [1]. Most of all it happens on the working surface of the machine and grinding bodies in zones of their contact with a crushed material.

All of these factors play an important role in the mill. An example of this is pressure and temperature. During grinding of the material of which the mill is constructed of the other substances add impurities to the mill. The quantity of Fe(me) in the activated oxide (MoO_3) increases with further activation time and after 90 min. we have 0,38% (table 1).

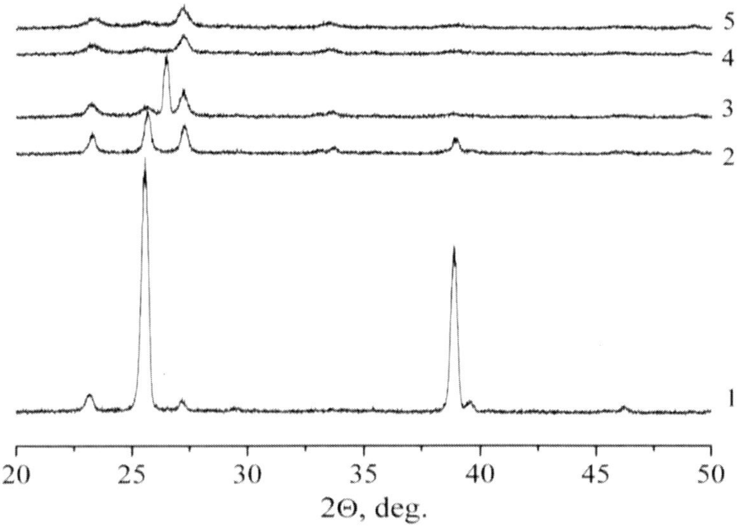

1- Original MoO_3 2 –15 mins of MA 3-30mins of MA 4 -60 mins of MA 5-90 mins of MA.

Figure 2. Changing of diffractograms during MA of molybdenum oxide MoO_3.

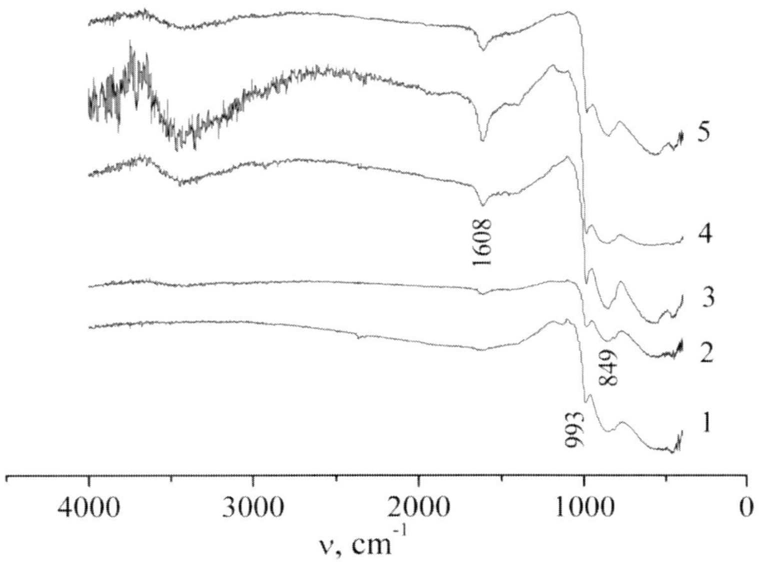

1- Original MoO_3 2 –15 mins of MA 3-30 mins of MA 4 -60 mins of MA 5-90 mins of MA.

Figure 3. Changing of IR- ranges of MoO_3 during MA.

Researches of samples by the X-ray method showed that initial MoO_3 has the following characteristics: D=340 Å; the size of microdeformations is 0,26%. As a result of mechanical activation within 90 min. the size of area of dispersion decreased to 220 Å, and the size of micro-deformations increased to 0,57% (table 1).

This shows that the result of mechanical activation, the integrated intensity reducing and broadening of characteristic reflexes of oxides of molybdenum, are caused by processes of crushing and amorphization (figure 2). According to infrared spectroscopy (figure 3) in samples of MoO_3, which had undergone mechanical activation we can see intensive absorption in area 1400 - 1600 cm^{-1}.

During mechanical activation we can see superficial interaction of the crushed material with vapors of water and CO_2 from the air.

During mechanical activation of α-Fe_2O_3 on sciagram we can see the gradual reduction of integrated intensity of reflexes caused by gradual amorphicity and partial restoration of hematite by the iron which is a part of the crushing material.

It is established that after 60 min. of MA the maintenance of Fe^{2+} makes 1,5%.

During application to a solid body of mechanical energy we can see how the material is redistributed. One part of energy is spent on formation of a new surface, part of energy is spent on formation of defects of crystal structure, chemical transformations, and the part passes into heat [2, 3, 4].

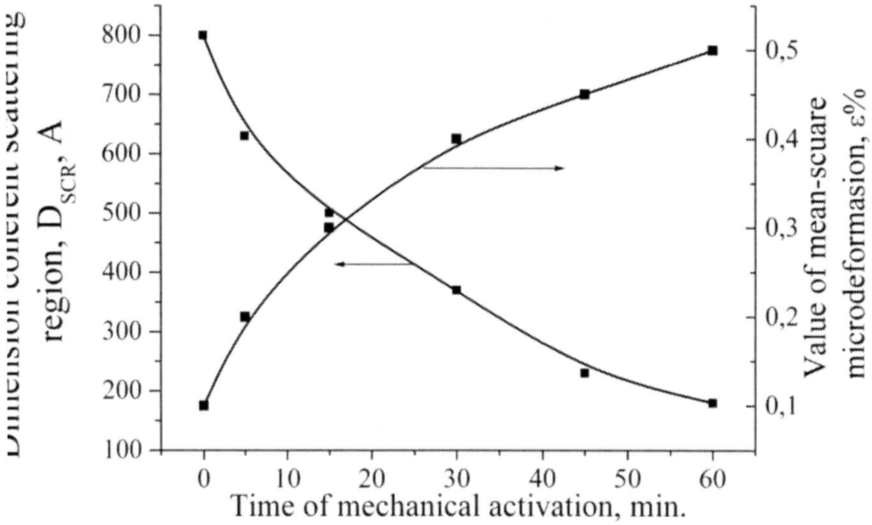

Figure 4. How Size of coherent dispersion and size of microdeformations depend on grinding time in a mill.

Part of the energy is spent on formation of a new surface; part on formation of defects of crystal structure and chemical transformations; and part changing into heat.

According to X-ray diffraction it is established that during MA the size of crystals getting smaller from 800 to 180 Å and micro-deformations increases from 0,1 to 0,5% (figure 4).

During mechanical activation of iron oxide it intensively reacts with vapors of water and carbon dioxide contained in the air. This leads to hydration and carbonization of the surface. According to infrared spectroscopy in samples of MoO_3 subject to mechanical activation we can see intensive absorption in the field of 1400 - 1600 cm^{-1}.

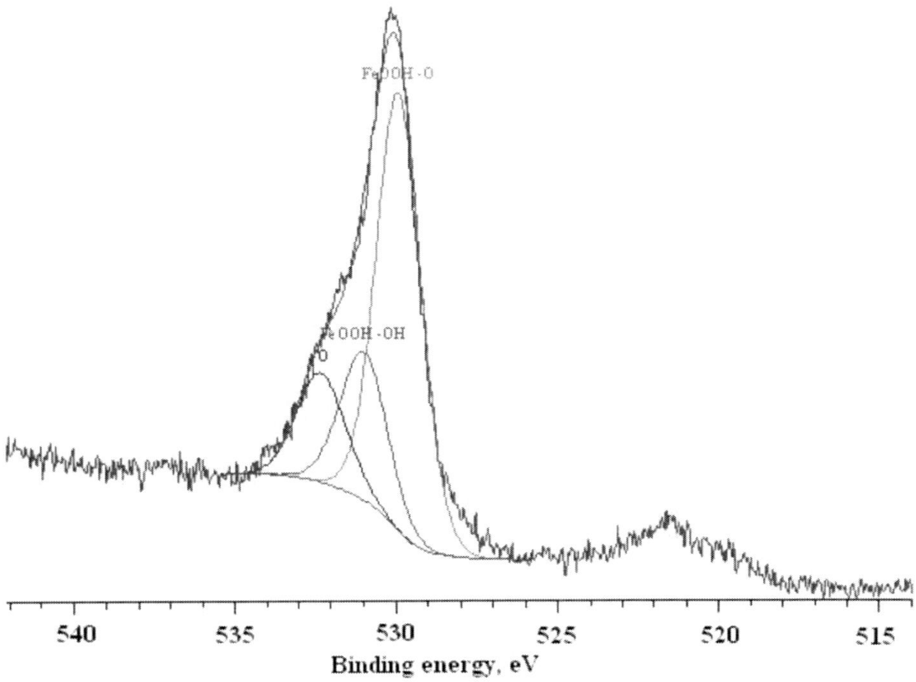

Figure 5. (a) 1Os Spectrum of hematite after 15 mins of MA.

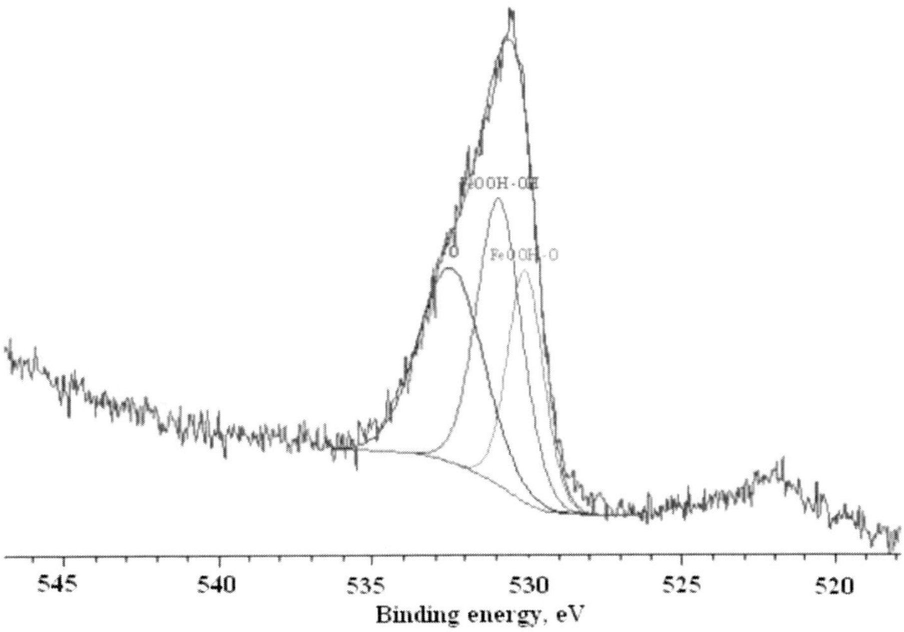

Figure 5. (b) 1Os Spectrum of hematite after 45 mins of MA.

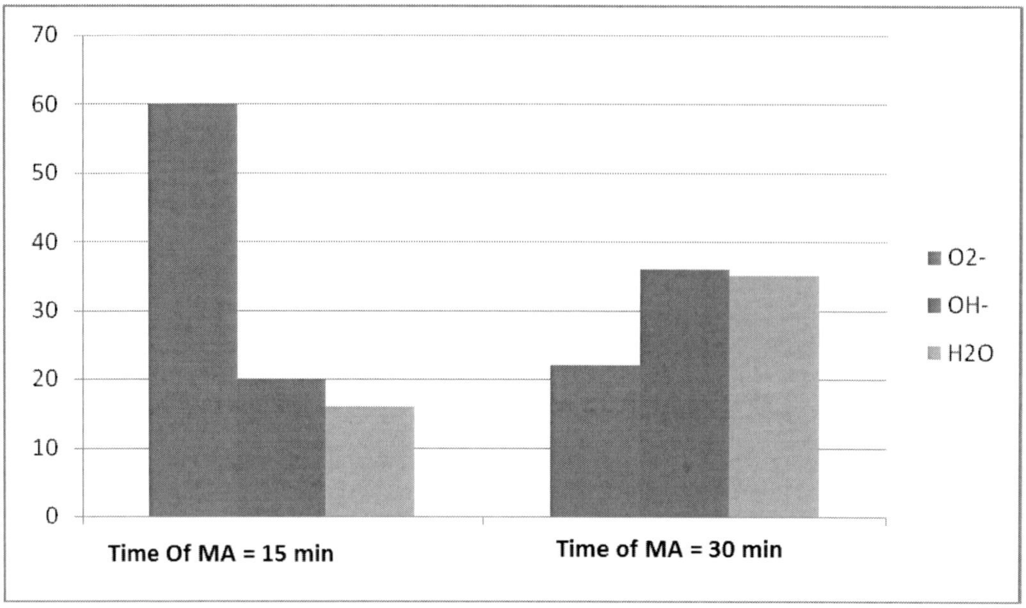

Figure 6. Various states of oxygen on a surface.

Research of the structure of the surface by X-ray photoelectron spectroscopy (figure 5 a,b) showed that iron on a surface is in a form α-FeOOH that corresponds to the energy 711.5 and will be coordinated with other data [5]. Oxygen has no less than 3 components. States which are characterized by O1s peaks with energy of communication is 529,9 eV and 530,90 eV belong to O_2-and OH-being a part of FeOOH [6]. Form of oxygen with a binding energy is 532.46 eV and can reliably be attributed to the associated surface water. If we increase time of mechanical activation, the quantity of water connected with a surface increases. (figure 6). Thus, During MA the surface area is increasing, the size of dispersion area decreasing and this increases the size of micro-deformations. At the same time crushing processes of secondary aggregation occur.

Mechanochemical Synthesis of Iron Molybdate

By the X-ray method it is established that in the process of mechanical activation of MoO_3 and α-Fe_2O_3 taken in a stoichiometric quantity 3:1 within 75 minutes of MA we can see reduction of characteristic reflexes of the MoO_3 phase and amorphicity of α-Fe_2O_3. We can see reduction in sizes of molybdenum oxide (VI) from 260 Å to 100 Å during 75 min of mechanical activation and the size of micro-deformations increase from 0.3 to 0.55 %.

During calcinating of samples we can observe changes in their structures and if we heat it up to 400 °C $Fe_2(MoO_4)_3$ starts to form. Increasing temperature of calcinating up to 500 °C leads to the reversible polymorphic transformation $Fe_2(MoO_4)_3 \leftrightarrow Fe_2(MoO_4)_3$ (figure 7), что подтверждается данными работы [7].

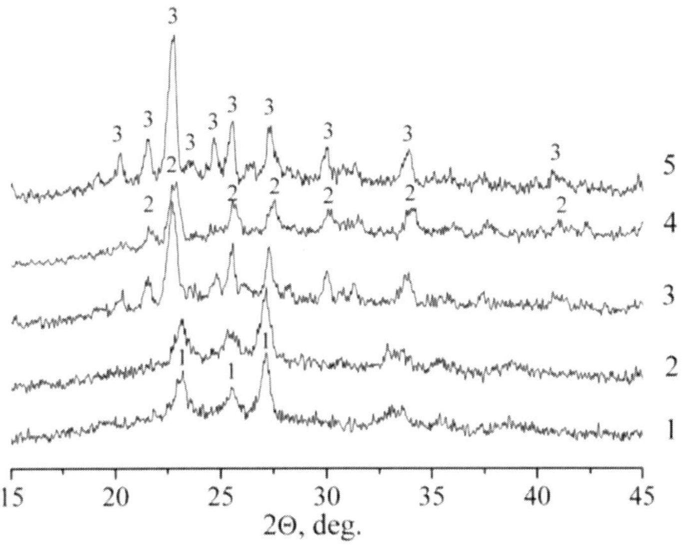

1- MoO_3, 2- $Fe_2(MoO_4)_{3monocl}$, 3- $Fe_2(MoO_4)_{3orthoromb.}$

Figure 7. X-ray diagram of α-Fe_2O_3 and MoO_3 depending on the calcination temperature.

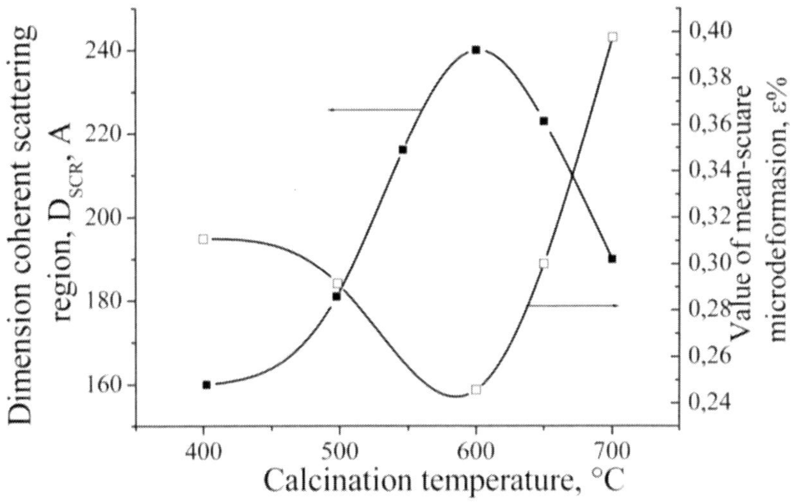

Figure 8. Changing in the size of the crystallites and micro deformations of $Fe_2(MoO_4)_3$, depending on the calcination temperature.

Temperature of calcination °C: 1-200 ; 2-300 ; 3 - 400; 4 - 500; 5-600

Thus calcination with temperatures over 400 °C increase average size of crystallites from 160 to 240 Å at 600 °C.

However if we increase the temperature of calcinating to more than 600 °C, we will notice a decrease in the size of crystallites to 180 Å at 700 °C.As a result of polymorphic transformations peaks get bigger, size of crystallites decreases and crystal lattice changes more rapidly During calcinating, microdeformations change their sizes. During heat treatment at 400 – 600 °C the size of microdeformations decreases from 0.31 to 0.24% (figure 8).

Calcinating at more high temperatures - 700 °C leads to an increase in the size of microdeformations to 0.4% - it is called polymorphic transformations.

Analysis of thermograms (figure 9) shows that calcination mixture of Fe_2O_3 and MoO_3 taken in a stoichiometric molar ratio of 1: 3 consists of three thermal effects.The first thermal effect – endothermic, lies in the temperature range 20 - 120 ° C occurs due to the removal of adsorbed moisture and CO_2.The second thermal effect - exothermic, lies in the temperature range 280 - 370°C blurred and obviously occurs due to the formation of structures of iron molybdate $Fe_2(MoO_4)_3$ and its polymorphic transitions from orthorhombic to monoclinic. Third, the exothermic heat effect, happens because of melting of iron molybdate.

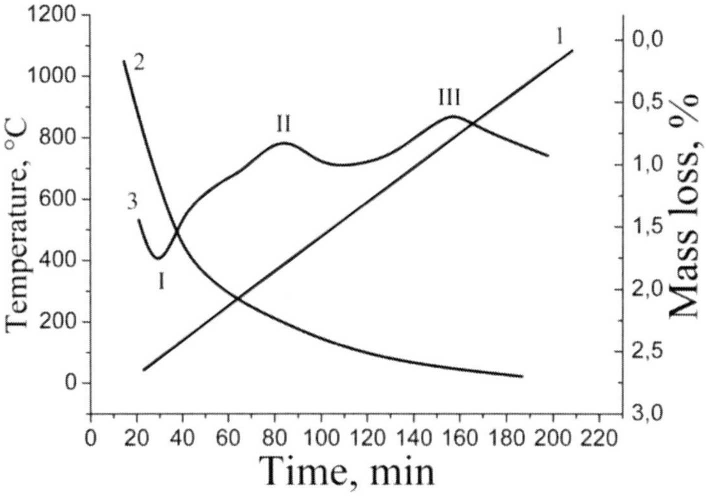

1-line of temperature
2-curved line
3 - curved line of DTA

Figure 9. Thermogram of MoO_3 and α-Fe_2O_3 after MA.

It is known that at methanol oxidation in formaldehyde the best catalyst is: surplus of MoO_3 in comparison to $Fe_2(MoO_4)_3$. Excess of molybdenum is necessary for maintenance of stoichiometric quality of the catalyst during the molybdenum ablation [8]. In an elementary cell of molybdate can enter without its distortion no more than 10 atoms of molybdenum, a nuclear ratio of Mo/Fe in molybdate is 2,1 [8]. Enter of excess molybdenum into a molybdate lattice by replacement ions of iron with ions of molybdenum is improbable as theoretical calculations of density of the replaced molybdate doesn't correspond to the experimental density [8]. Therefore for research of catalytic properties of iron molybdate with excess of Mo, we prepared samples of catalysts with a nuclear ratio of Mo/Fe=2,1 owing to activation of Fe2O3 with excess quantity of MoO_3 in a mill during 45 minutes and calcination at 400 °C lasting for 6 hours.

It should be noted that the specific surface of iron molybdate is 28 m2 / g which is 2.8 times more than the specific surface of iron molybdate.

MECHANOCHEMICAL SYNTHESIS OF CALCIUM MOLYBDATE

Mechanochemical synthesis of calcium molybdate is done by mechanical activation of $Ca(OH)_2$ and $(NH_4)_6Mo_7O_{24}\cdot 4H_2O$.

By the X-ray method it was found that during the starting moments (0-5 min.) of mechanical activation, the formation of crystalline calcium molybdate was observed (figure 10).

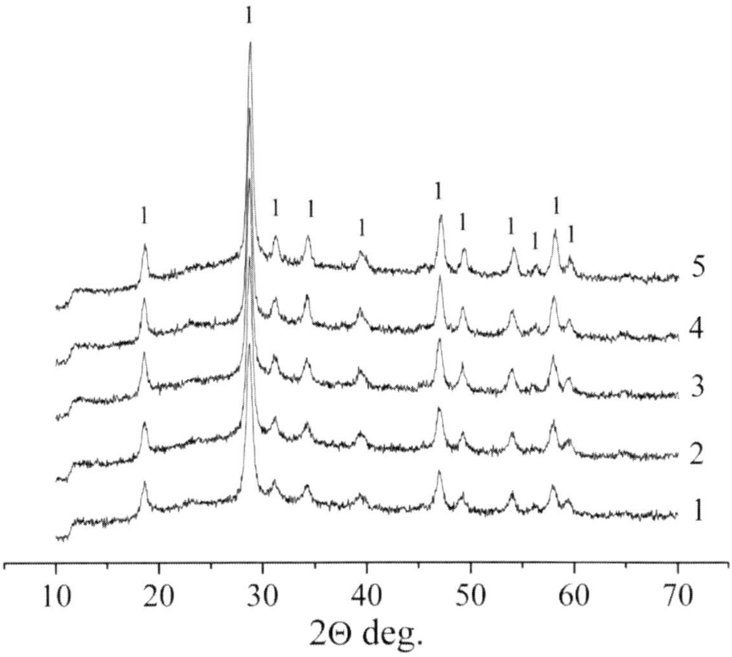

1-$CaMoO_4$
Time of MA, min.: 1-5, 2-15, 3-30, 4-45, 5-60.

Figure 10. X-ray diagram of $Ca(OH)_2$ and $(NH_4)_6Mo_7O_{24}\cdot 4H_2O$.

Calculation of the parameters of the crystal structure revealed that with increasing time of MA we can see how the crystallite size of $CaMoO_4$ increased to 146 Å after 5 min. of mechanical activation and up to 172 Å after 60 minutes and the microdeformations decreased from 0.89 to 0.75% (figure 11). Probably during initial stages of mechanical activation there is an accumulation of defects of crystal structure and intensive interaction between $(NH_4)_6Mo_7O_{24}\cdot 4H_2O$ and $Ca(OH)_2$ on the active centers of a surface and later the reduction is observed.

We found out that during mechanical activation of molybdenum oxide there is an increase in size of a specific surface, reduction of the size of coherent area with increase in sizes of microdeformations. At the same time the crushing process of secondary aggregation proceeds. During interaction of grinding bodies with a crushed material, the part of material of which the mill is made from, passes into activated substance from its working surface thus polluting it. MA leads to interaction of a crushed material with air leading to carbonization and hydration of the surface.

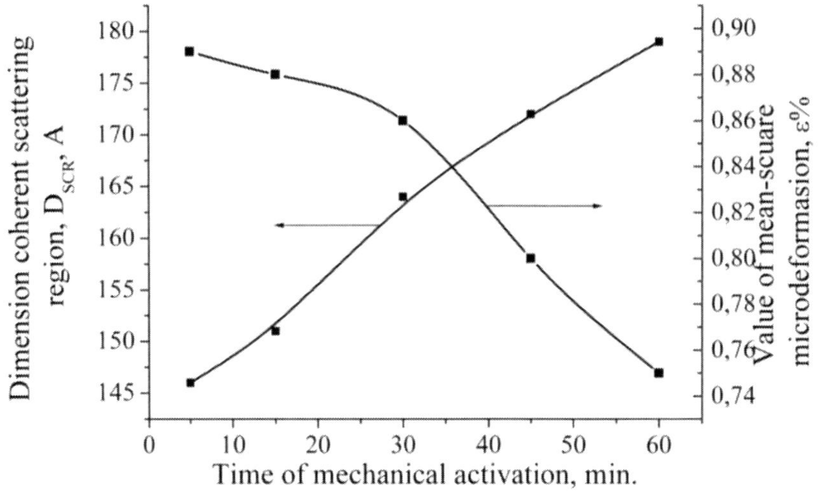

Figure 11. Dependence of average crystal sizes of $CaMoO_4$ from MA time.

During MA of iron oxide also there is a reduction in the sizes of crystallites and increase in size of microdeformations.

It is established that iron molybdate is formed at a stage of heat treatment at a temperature of 400-500 °C. Besides, during of heat treatment of iron molybdate there is a polymorphic transition: $Fe_2(MoO_4)_3 \leftrightarrow Fe_2(MoO_4)_3$ Synthesis of calcium molybdate happens during mechanical activation of $Ca(OH)_2$ and $(NH_4)_6Mo_7O_{24}*4H_2O$.

Research of Catalytic Properties of Molybdenum Oxide and Molybdates of Fe, Ca in the Oxidation of Methanol in Formaldehyde Have Been Done

During many years the main production method of receiving formaldehyde around the world was catalytic interaction of methanol with oxygen in air.

Now we can produce formaldehyde by oxidation of methanol on catalysts: at a temperature of 600 – 700°C on the silver catalyst or with excess of air on iron molybdate at a temperature of 250-400°C.

During oxidation of methanol on oxidic catalysts proceeds as the following reactions:

$$CH_3OH + 2MoO_3 \rightarrow CH_2O + H2O + Mo_2O_5$$
$$Mo_2O_5 + 1/2O_2 \rightarrow 2MoO_3$$
$$CH_3OH + 1/2O_2 \rightarrow CH_2O + H_2O$$

Main part of transformation of methanol on oxidic contact, as well as on the silver catalyst, is reaction of oxidizing dehydrogenation of methanol.

The only unimportant reaction of this process is further oxidation of the formed formaldehyde:

$$CH_3OH + O_2 \rightarrow CH_2O \xrightarrow{O_2} HCOOH \xrightarrow{O_2} CO_2 + H_2O$$

Researches of catalytic model samples were conducted on installation of flowing type, in reaction of oxidizing dehydrogenation of methanol in formaldehyde in the range of temperatures of 200 – 400°C. Important influence on quantity of target product – formaldehyde, is rendered by temperature of the process, and also the speed of gas. During the experiment on the industrial catalyst we found out that the volume speed in experimental conditions is 17000 h^{-1}. At the volume speed of 8500 h^{-1} productivity of formaldehyde at 300°C is 11,39. The increase in volume speed to 17000 h^{-1} leads increase in productivity to 13,87 (figure 12). The further increase in volume speeds leads to even bigger decline in production of formaldehyde owing to oxidation of the formed methanol to formic acid, carbon and water dioxide.

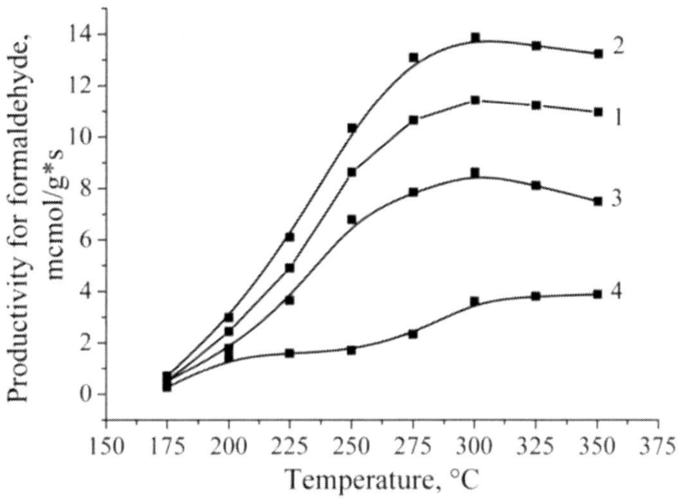

Speed of gas:1 – 8500, 2 – 17000, 3 – 42500, 4 – 85000.

Figure 12. Dependence of formaldehyde by weight of the catalyst sample.

It was established that molybdenum oxide has a low catalytic activity, and its performance on formaldehyde is 0.3micro mol/g*s. It is established that Mechanical activation leads to growth of catalytic activity in the initial stage of activation (15-45 min.). For example, samples activated within 45 min. show that their productivity increases by 1,5 times in comparison with initial molybdenum oxide molybdenum(0,45micro mol/g*s). The increase in activity is explained by growth of size of specific surface. Besides, in system there is an accumulation of metal iron which owing to mechanical activation partially gets oxidized to iron oxide which interacts with molybdenum oxide leading to the insignificant quantity of amorphous $Fe_2(MoO_4)_3$ that is formed. Decrease in activity during increase in time of mechanical activation happens because of decrease in surface area from 55 to 18 m^2/g. Catalytic activity of calcium molybdate was estimated owing to size of specific productivity of the microreactor. It is established that catalytic activity of calcium molybdate increase in range of temperatures 250 – 340 °C from 6 to 7,2 micro mol/g*c (table 2). It should be noted that in the received product there are no impurity of formic acid, and there is no CO_2 in the gas phase that shows high selectivity of molybdates. The further increase in temperature of process to 400 °C leads to oxidation of the formed formaldehyde to CO_2 and H_2O and decline in production on formaldehyde to 6,5 micro mol/g*c (table 2).

Table 2. Dependence of catalytic activity of calcium molybdate on temperature

Temperature, °C	180	210	240	270	300	340	400
Productivity for formaldehyde, mcmol/g*s	0,3	4,8	6	6,3	6,8	7,2	6,5

With increasing of temperature, the catalytic activity and selectivity iron molybdate of formaldehyde is monotonically increasing and reaches its maximum at 300 – 320 °C. Further temperature increase leads to growth of conversion of methanol, but at the same time reduces activity for the target product. Also there is a formation of other products of synthesis, with total selectivity falling down to 92,5% at 400°C. Owing to special method it was established that there is an insignificant amount of acetaldehyde and methylacetate. Calcium molybdate showed good results on activity and selectivity. If we add small amounts of calcium oxide we can make catalysts more strong. It is proved that if we add calcium molybdate to our catalysts, their granules will be stronger, however, along with it there is a reduction of a specific surface (table 3). Results of testing catalytic activity showed that it we add 2% of calcium molybdate we will not notice big change in activity of the catalyst, however further increase in quantity of $CaMoO_4$ leads to more considerable changes of activity.

Table 3. Meaning of yield strength and specific surface of samples

	Prototype [9]	Sample without $CaMoO_4$	Sample with $CaMoO_4$, %			
			2	4	6	8
Durability, MPa	6,0	10,1	12,2	17,5	20,1	26
Surface area, m^2/g	14	28,4	25	20	18	13

Thus, adding 2% of $CaMoO_4$ allows to increase catalyst durability without reducing its activity.

Studying of the synthesis processes and catalytic properties of iron and calcium molybdates was supported by the Fund Assistance to Small Innovative Enterprises in Science and Technology (contest UMNIK-2014).

Studies on the effect of mechanical activation on the structure of iron oxide and molybdenum were made in the framework of the state assignment program. Project №1800.

REFERENCES

[1] A.A. Il'in, N.N. Smirnov, A.P Il'in // *Izv. VUZov. Himija i him. tehn.,* 2005. – T.48. – Vyp.6. – S.83 - 87.

[2] Molchanov, V.V. Nauchnye osnovy primenenija metodov mehanohimii v prigotovlenii katalizatorov / V.V. Molchanov, R.A. Bujanov // *Tez. dokl.* 4 Rossijskoj konferencii: Nauchnye osnovy prigotovlenija i tehnologii katalizatorov. – Sterlitamak, 2000. – P.48.

[3] V.V. Boldyrev, E.G. Avvakumov Mehanohimija neorganicheskih veshhestv // *Zhurn. Uspehi himii,* 1971. – T. 40. – P.1835-1856.

[4] R.N. Rumyantsev, A.A. Il'in, A.P. Il'in, I.V. Pazukhin // *Theoretical and Experimental Chemistry,* 2011 – T.47. – №1, P.41-44.

[5] A. P. Grosvenor, B. A. Kobe, M. C. Biesinger et al. // *Surf.Interface Anal,* 2004. – №36. – P.1564-1574.

[6] Analiz poverhnosti metodami Ozhe- i rentgenfotojelektronnoj spektroskopii / *pod red. D. Briggsa i M Siha. – M: Mir,* 1987. – 600 p.

[7] L.M. Pljasova // *Journal of Structural Chemistry,* 1976. – T. 17. – № 4. – P.637-640

[8] B.I Popov, L.I. Abaulina, G.N. Kustova // // *DAN SSSR*, 1975. – T.221. – № 6. – P.1359-1362.

[9] E. Konka, K. Rubini, M. Marki Method of preparing catalyst for oxidising methanol to formaldehyde // Pat. RU №2388536, 2010.

INDEX

D

G

H

I

N

O

P

Q

R